He who has an ear, let him hear what the Spirit says to the churches. To him who overcomes I will give to eat from the tree of life, which is in the midst of the Paradise of God.

Revelation 2:7

PREFACE

Purpose of the Book

This book is written for fellow mathematicians who, as purveyors of "the language in which God has written the Universe" (Galileo Galilei, 1564 –1642), may harbor a deep yearning to know more about this God and wonder whether the language itself could reveal His will or plan for them, and be used to develop a basis for His worship. To use mathematics for this purpose requires additional axioms reflecting the beliefs that uphold this God.

Over the span of their professional life, mathematicians greatly enjoy the power of their minds that are unconstrained by the laws of nature as they methodically traverse one space or universe after another, creating the simplest non-natural mathematical entities such as negative real numbers to the most sophisticated ones that defy physical laws such as Zermelo's Axiom of Choice. Mathematicians are excited at the prospect of discovering and conquering a new universe, and fearless, persistent and unruffled as they set the rules for this new universe by their out-of-this-world axioms. They rub their hands as their mathematical universe unfolds itself before their very eyes, breathtakingly consistent and beautiful.

Uncharacteristically, however, mathematicians appear disinterested in, if not unnerved by what arguably is the last frontier to be conquered – religion! And, perhaps, understandably so, given that religion is a vast and complex body of knowledge that is based on faith in a supernatural being. Nonetheless, the authors argue that if the supernatural being is a non-natural entity, then the only fighting chance that humankind has to know more about this incorporeal being via a scientific endeavor is mathematics. Among the scientific disciplines, mathematics is the only one that can build a solid bridge of rigor between science and religion. After all, mathematics has the greatest power derived solely from the human creative mind to create what has never before existed such as the complex number $i^2 = -1$. Mathematics leads, discoveries follow!

Indeed, as the authors show in this book, mathematics, to begin with, can determine whether a set of beliefs in a religion is consistent within an axiomatic framework that reflects the core assumptions or accepted truths of this religion. The authors consider Christianity because with the advent of the Internet there is so much readily available information about the religion and Judaism – not to mention the abundance of freely available online bible study resources – that it is now possible for mathematicians and other scientists and theologians to meaningfully work together to begin to study the Bible with a new tool of biblical interpretation – what we call *Biblical Mathematics*. The purpose of biblical interpretation is to know what God the Father has willed for each and every individual who believes in His Son, Jesus Christ. Surely the ultimate goal in life of every God-fearing individual is to know the revealed will of God for him or her, whether through an act of faith or through a systematic scientific mode such as mathematics.

In this book, the authors develop a first-ever mathematical framework, which includes a new axiom, for a new applied field called Biblical Mathematics, within which one can study the Bible in a systematic manner. Though the term "Biblical Mathematics" has been used before, it has not been defined precisely.

The purpose of the book is to show the consistency of results derived within the framework of biblical mathematics. This framework is discussed in detail in Chapter 1.

Goal and Objective of the Book

To achieve the purpose of the book, the goal of the book is to answer two questions pertaining to Ephesians 2:8 – 9, quoted below from King James Version:

> [8] For by grace are ye saved through faith; and that not of yourselves: it is the gift of God: [9] Not of works, lest any man should boast.

These questions are:

1. What is precisely the *faith* referred to in Ephesians 2:8?
2. Given that one's deed is not required to receive God's gift of grace, *how* then do we *say* or *proclaim* this faith and *to whom* should we proclaim this faith?

The objective of the book is to develop and use the methods of biblical mathematics to achieve the goal of the book. The methods are discussed in Chapter 1.

Outcomes of the Book

It is shown that an outcome of biblical mathematics is the astonishing conclusion that the Lord's Prayer itself is the proclamation of faith mentioned in Ephesians 2:8 and the seemingly unremarkable number 153 in John 21:11 links the Lord's Prayer with our faith in the fulfillment if the will of the Father in His Son, Jesus Christ. Apart from providing the answers to (1) and (2), biblical mathematics also shows that the Lord's Prayer is actually a deep reservoir of messages that are the central tenets of the Christian faith and is a means to partake in sanctification. Chapters 2 to 4 expound on these.

What is more astonishing though is the unexpected result that there exist daily prayer times of the Lord's Prayer derived from the permutations of the digits of the numeral 153, and that the prayer times represent the last 7 hours of the agony of Jesus Christ on the cross and include the precise time of His death and the precise time when His side was pierced. The details are provided in Chapters 5 – 7.

The final result, detailed in Chapter 8, is totally unanticipated: *the Lord's Prayer is a covenant that Jesus Christ made with those who believe in Him – it is the covenant of the Redeemer of Isaiah 59!*

The answer to the question of what the will of God is for us – the mathematicians and scientists – is now clear. If we accept the outcomes of the book, then we have found our creed, the Lord's Prayer, taught to us by no sinful mortal, but by God Himself! We now know how to communicate with and worship Him and what His promises are for us when we end our earthly lives. Therefore, the will of God for us is to discover and share with our fellow human beings this profound knowledge and subsequent knowledge about Him via our beloved profession, and in the process, glorify Him.

Bible Study Resources

Bible Version
The verses of the Bible are taken from King James Version (KJV). However, for the sake of clarity, other versions are also used, in which case, the versions are indicated.

Hebrew and Greek Texts
The main online Bible study resources used are www.biblegateway.com, www.scripture4all.org and biblehub.com, accessed for the original Hebrew and Greek texts and their meanings, transliterations, interlinear translations, lexicons and word origins using Strong's Concordance, Englishman's Concordance, NAS Exhaustive Concordance, the Brown-Driver-Briggs Hebrew and English Lexicon of the Old Testament and the Thayer's Greek Lexicon.

Biblical Terms and Expressions
The meanings of biblical terms and expressions are sourced from the King James Bible Dictionary accessible at www.kingjamesbibledictionary.com.

Acknowledgment

The authors acknowledge the Holy Spirit who they believe imparts the knowledge of God and of His Son, Jesus Christ to those who believe in Jesus Christ.

The authors acknowledge their immediate families for their support and encouragement and to all the members of the Fiji House of Prayer, including Team 19861, for their prayers.

The principal author, Jito Vanualailai, dedicates this book to his beautiful wife Nola, and wonderful children Maria, Tito and Caroline, and to his lovely nieces Taina and Evelyn.

Jito Vanualailai
Eroni Tomasi
Paulo Vanualailai
Jope Takala
Suva, Fiji
1 July 2018

ABOUT THE AUTHORS

Jito Vanualailai was born in the village of Fatima on Rabi Island, Fiji, on 1 July 1965. He attended the University of the South Pacific, graduating in 1987 with the Bachelor of Science degree mathematics and physics majors. In 1988, he was awarded a Japanese Government scholarship to undertake postgraduate studies in Japan. In 1990 he graduated with the Master of Engineering specializing in Control and Systems Engineering from the University of the Ryukyus, Okinawa. In 1991 he was accepted by Kobe University, Kobe, Japan, to undertake the PhD program in Mathematical Control Theory. He successfully completed the program in 1993 with the publication of his PhD thesis *Applications of the Direct Method of Lyapunov to Nonlinear Systems* and returned to Fiji where he joined the University of the South Pacific (USP) in 1994 as Lecturer in Mathematics. He has been with USP since then, and is now Professor of Applied Mathematics. Since 2011, he has been the substantial Director of Research at USP. Professor Jito Vanualailai has published extensively in his field over the last 24 years as an academic and researcher.

Eroni Tomasi was born in the village of Drekeniwai, *Tikina* of Navatu, in the province of Cakaudrove, Fiji, on 24 July 1971. He studied law for a year (in 2000) at the University of the South Pacific before he decided to pursue evangelism, his life's true calling, and joined the Scripture Union Fiji as a volunteer, inspiring children and young people in schools in Cakaudrove to know God. In 2002, he founded the Fiji House of Prayer (FHOP) which is an inter-denominational evangelical Christian movement that welcomes anyone who desires a life of continual prayer and worship. With the membership of the FHOP growing exponentially over the years, Apostle Tomasi has begun the establishment of a five-fold ministry, tentatively called the Aleph – Tav Ministries Fiji, planned to be operational by the end of 2018.

Paulo Vanualailai was born in the village of Rakentai on Rabi Island, Fiji, on 23 May 1963. From 1984 – 1987, he attended the Catholic's Pacific Regional Seminary where he studied Christian Theology. In 1989, he enrolled at the University of the South Pacific in the Bachelor of Arts program, from which he graduated in 1993 majoring in physical geography. He worked as a secondary school teacher from 1994 – 1995. From 1996 to 1998 he worked for the Fiji Government as the first-ever Coordinator of Climate Change. Subsequently he was successful in securing a Japanese Government scholarship to undertake postgraduate studies at Ibaraki University, Japan. He successfully completed the MSc in Coastal Engineering in 2002 and the PhD in the same field in 2005 with the publication of his PhD thesis *Assessment of Coastal Protection Systems in the Pacific*. On his return to Fiji, he had a stint with the World Wildlife Fund as a Sustainable Tourism Adviser in 2005, and with the University of the South Pacific as Lecturer in Geography in 2006. Dr. Paulo Vanualailai is now a successful entrepreneur, having moved out of the public sector in 2007 and created his consultancy firm Envi-Green Fiji Ltd which specializes in environmental impact assessments. Over the years he has conducted comprehensive research on the impact of climate change in the Pacific Island Countries and published many reports for his clients including those in the private sector, governments and the tourism industry.

Jope Takala was born in the village of Urata, *Tikina* of Wailevu in the province of Cakaudrove, Fiji, on 1 January 1979. He graduated from the University of the South Pacific in 2001 with the Bachelor of Education majoring in education and mathematics. He taught mathematics for ten years at several secondary schools in Fiji before joining the School of Computing, Information System and Mathematical Science of the University of the South Pacific in July 2011 as Teaching Assistant in Mathematics.

Table of Contents

PREFACE 2
   Purpose of the Book 2
   Goal and Objective of the Book 4
   Outcomes of the Book 5
   Bible Study Resources 6
   Acknowledgment 7
ABOUT THE AUTHORS 8
CHAPTER 1: Biblical Mathematics 13
   Framework 13
   Methods 17
      Method of the Gematria 17
      Biblical Numerology 20
      Method of Verse Identification 21
   Summary of Results 27
CHAPTER 2: The Number 153 and the Will of the Father 31
CHAPTER 3: The Lord's Prayer and the Will of the Father 38
   Sovereignty of the Father 45
   Fulfillment of the Will of the Father 46
   Acknowledgment of the Father 48
   Becoming More Like Jesus Christ 51
   Fullness of Jesus Christ 54
CHAPTER 4: The Lord's Prayer and Our Sanctification 59
   The Lord's Prayer 61

Jesus Christ is the Judge of Humanity 64
CHAPTER 5: The Lord's Prayer and the Lord's Last Prayer 66
CHAPTER 6: The Lord's Prayer and the Death of Jesus Christ 74
CHAPTER 7: The Lord's Prayer and Prayer Times 85
CHAPTER 8: The Lord's Prayer is the Covenant of Jesus Christ .. 93
    Pledges 94
    Signs 94
APPENDIX 101
Hebrew Gematriot 101
Greek Gematriot 103
Tables of Gematriot 105
  Tables for Chapter 1 105
  Tables for Chapter 3 106
    Sovereignty of the Father 107
    Acknowledgment of the Father 107
    Becoming More Like Jesus Christ 110
  Tables for Chapter 6 111
  Tables for Chapter 7 112
Isaiah 53 113

CHAPTER 1: Biblical Mathematics

Framework

Christianity is based on the teachings of Jesus Christ, believed to have lived about 2000 years ago in the Holy Land. His teachings are recorded in the Bible, the holy book for Christians for whom the truth is captured in one single verse of the Bible, namely verse 16 of the third chapter of the Gospel of John (John 3:16, King James Version):

> ^{16} For God so loved the world, that he gave his only begotten Son, that whosoever believeth in him should not perish, but have everlasting life.

An axiom must reflect this truth. We propose the following statement (Axiom 1.1) underpinning this core belief that is also expounded in the First Epistle of John (1 John). It is taken from the 4th chapter (1 John 4). That Jesus existed historically is beyond doubt, given the voluminous supporting literature by the majority of modern scholars of antiquity on the historicity of Jesus. For billions of Christians, this is an undeniable fact. To recognize this truth, we exercise the power of mathematics – we create and we work consistently within a mathematical framework propped up by an axiom, proposed as follows:

Axiom 1.1 *Jesus Christ came in the flesh.*

The First Epistle of John was written to counter the doctrine of *Docetism*, a belief among some early Christians (AD 95 – 110) that *the humanity of Christ, his sufferings, and his death were apparent rather than real* (Collins English Dictionary).

The 4th chapter addresses Docetism (1 John 4: 1 – 3, New KJV):

> ⁴ Beloved, do not believe every spirit, but test the spirits, whether they are of God; because many false prophets have gone out into the world. ² By this you know the Spirit of God: Every spirit that confesses that Jesus Christ has come in the flesh is of God, ³ and every spirit that does not confess that Jesus Christ has come in the flesh is not of God. And this is the *spirit* of the Antichrist, which you have heard was coming, and is now already in the world.

God is love, and therefore *sent his Son to be the propitiation for our sins* (1 John 4:10). Hence the acceptance of Axiom 1.1 implies the acceptance of John 3:16, and vice versa.

To mathematicians, the representative statement of Axiom 1:1 is in fact the most convenient of propositions, for by first-order logic, we can easily formalize it as follows:

$$\exists X: X = \text{JESUS CHRIST} \cap X = \text{MAN}$$

With the adoption of this additional axiom, we create a new branch of mathematics called *biblical mathematics*.

Definition 1.2 *Biblical Mathematics is the application of mathematical methods to the study of the Bible, using the existing mathematical axioms, Axiom 1.1 and the scriptures themselves.*

Though the term "Biblical Mathematics" has been used before[1], it has not been defined precisely.

The terms and expressions in biblical mathematics necessarily include biblical terms and expressions. For their meanings, one can refer to readily available bible dictionaries, for instance, the King James Bible Dictionary accessible at www.kingjamesbibledictionary.com.

A list of mathematical axioms and their descriptions can be found at mathworld.wolfram.com/topics/Axioms.html.

Biblical mathematics can be a method of interpreting the Bible. It can help us see biblical messages in a new light, from a new perspective. It can further enhance our understanding of the biblical God, and by extension, therefore, His Son, Jesus Christ, and His Holy Spirit. Like any other branch of mathematics, biblical mathematics must create new knowledge within its axiomatic structure. However, unlike other branches of mathematics, we can actually prescribe how the new knowledge should be used by invoking Deuteronomy 29:29 (King James Version, KJV):

> ^{29} The secret things belong to the LORD our God, but the things revealed belong to us and to our children forever, that we may follow all the words of this law.

Our discoveries, the things revealed to us by God, should lead us to be obedient to God, for they constitute His revealed will

---

[1] Ed F. Vallowe, *Biblical Mathematics: Keys to Scripture Numerics*, Ed F. Evangelist Society, 1984

for us and is that which we must carry out. We can demand, for there is nothing stopping us from doing so, that this be the outcome of biblical mathematics; that we specify the impact of biblical mathematics even *before* we derived its results. By doing so, we can have an elegant framework of biblical mathematic as shown in Fig.1.1, wherein we accept or reject a set of related results (definitions, lemmas, theorems, corollaries, conjectures) depending on whether the set is aligned to 1 John 4 and Deuteronomy 29:29. So as we unravel the Bible mathematically, we can take comfort in the fact that 1 John 4 and Deuteronomy 29:29 guide our work. Any mathematical result outside of this framework is not acceptable.

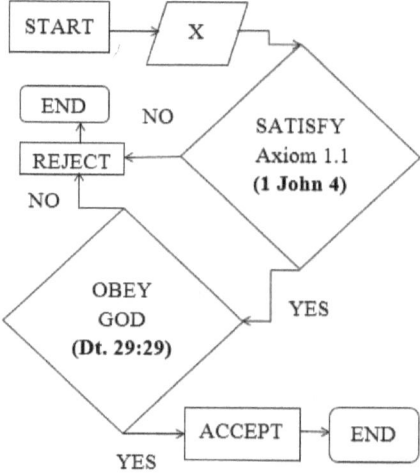

Fig. 1.1 The framework of Biblical Mathematics, where X is some result of biblical mathematics. If X is derivable from Axiom 1.1 (1 John 4) or in agreement with Axiom 1.1 and it can lead to obeying God (Deuteronomy 29.29), then it is acceptable.

Methods

Since biblical mathematics is a new area, its methodologies are waiting to be discovered. However, there are two known methods that have been used to interpret the Bible using the Basic Axioms of Algebra (equality, addition, subtraction and multiplication).

Method of the Gematria

Firstly, there is an ancient, but well-known and well-tested method used by the Jews to study and interpret the Torah (the five books of Moses) and the Tanakh (the Jewish Bible). It is called *gematria*, which is a system of assigning a numerical value to each letter in an alphabet. The Jews and Greeks have their own *gematriot or gematrias* (plural of gematria) with well-accepted *standard values*. These are listed in the Appendix. The Hebrew gematria has an extra set of values called the *sofit values*. These are alternate values for five existing letters that can appear at the end of a word. They are written differently but retain their pronunciation. Sofit values are also listed in the Appendix. In keeping with the traditional usage of the word "gematria", we shall refer to the value of a Hebrew or Greek letter as gematria.

For our purpose, we need to precisely define the method of gematria.

Definition 1.3 (Method of the Gematria) *Let V_1 and V_2 be two verses in the Bible. Let N_1 be the total gematria of a set of words in V_1. Let N_2 be the total gematria of a set of words in V_2. It is possible that V_1 and V_2 are contextually or thematically linked if and only if $N_1 = N_2$.*

Definition 1.3 refers to words or expressions in the verses that are contextual or thematic. By Collins English Dictionary, the *context* of a word, sentence, or text consists of the words, sentences, or text before and after it which help to make its meaning clear. So once the Method of the Gematria establishes the association ($N_1=N_2$) between the two verses V_1 and V_2, then several verses before and after them may need to be consulted for a better understanding of the association. Also, by the same dictionary, in linguistics, the adjective *thematic* denotes a word that is the theme of a sentence. In grammar, it is of or relating to the stem or root of a word. Very often, there is a need to consult Strong's Concordance and the Hebrew and Greek lexicons to fully understand the association between words that are thematic.

How will the gematria be used in biblical mathematics?

As an implication of Axiom 1.1, we accept that biblical writings are inspired by God. Indeed, in John 1:1 – 2, we read (KJV):

> [1] In the beginning was the Word, and the Word was with God, and the Word was God. [2] The same was in the beginning with God.

And in verse 14, we read:

> [14] And the Word was made flesh, and dwelt among us, (and we beheld his glory, the glory as of the only begotten of the Father,) full of grace and truth.

Now, "the Word" is clearly Jesus' title. In Greek, "word" is λογος, *logos*, which according to Strong's Concordance, is linked to the verb λέγω, *lego*, to bring a message to closure, or to say (speak), moving to a conclusion. So, "the Word" is also either the *written message of God*, or the *spoken word of God*.

By 2 Timothy 3:16 – 17 (New KJV), the Bible can therefore be used for instruction in righteousness:

> [16] All Scripture *is* given by inspiration of God, and *is* profitable for doctrine, for reproof, for correction, for instruction in righteousness, [17] that the man of God may be complete, thoroughly equipped for every good work.

With the availability of powerful personal computers, gematria has become an efficient and practical method of finding associations between biblical passages and verses. Because the mathematical equality is an association between two entities, gematria provides a means to see if there are meaningful associations between biblical messages. If there are, then we have a richer understanding of God and His will for us, which is the ultimate purpose of biblical interpretation.

Gematria is essential to Kabbalah, an ancient Jewish esoteric discipline that probes deep into Jewish sacred texts – the Torah and the Tanakh – for a better understanding of God, the universe and life. The Kabbalists believe that God created the universe through the power of the Hebrew letters and their numerical values. There is a mistaken view by some that Kabbalists use gematria for divination, perhaps because Kabbalah also delves into the mystical aspects of these texts. As shown in the Appendix, gematria is nothing but an alphanumeric system. Its application is governed within the Kabbalah framework expounded in *Zohar*, the chief text of the Kabbalah.

Similarly, within the framework of biblical mathematics (Fig.1.1), divination by the application of gematria or numerology has no foothold.

In this book, the gematriot of the original Hebrew and Greek texts are calculated using the alphanumeric codes given in the Appendix. They are checked against the Full Text Hebrew/Greek Bible Gematria Database developed by Richard A. McGough and accessible at his website www.biblewheel.com.

An example of the Method of the Gematria is provided at the end of this chapter (Example 1.2).

Biblical Numerology

Secondly, there is the more recent method of attaching a meaning to a number based on one or more coinciding events in the Bible. It is the method of *biblical numerology*. Some of the numbers themselves appear explicitly in the Bible. Such numbers are called *biblical numbers*. A first systematic study of biblical numbers is often attributed to the Harvard scholar Ivan Panin[2] (1855 – 1942). Several notable studies on the biblical meaning of numbers were made by E. W. Bullinger[3] (1837 – 1913) and Ed F. Vallowe[4] (1919 – 2002). A relatively

---

[2] I. Panin. *The Structure of the Bible: A Proof of the Verbal Inspiration of Scripture*, Gospel of Christ Print, 1891.
[3] E. W. Bullinger. *Number in Scripture: Its Supernatural Design and Spiritual Significance*, 4th Ed., Eyre & Spottiswoode, London, 1921.
[4] Ed F. Vallowe, *Biblical Mathematics: Keys to Scripture Numerics*, Ed F. Evangelist Society, 1984

recent book by Stephen E. Jones[5] is an excellent reference which builds on the works of Ed F. Vallowe and E. W. Bullinger. In this book, when we refer to biblical numbers, we shall refer to those defined and expounded in S. E. Jones' 2008 publication "The Biblical Meaning of Numbers from One to Forty."

Within the framework of biblical mathematics, we shall derive results using gematria and biblical numerology.

We now propose two new methods that depend on the location of verses in the Bible. We call these the *Method of Verse Identification* and the *Method of the Prime.*

Method of Verse Identification
When one shares a bible verse with others, one quotes its location in the Bible, identified by the name of the book, the chapter number and the verse number. Its location is synonymous with the message it carries. Thus when one says, for instance, "According to Genesis 1:1, God created the universe", one links the location "Genesis 1:1" with the actual message it carries, namely, "In the beginning, God created the heavens and the earth."

Chapter numbers and verse numbers are not unique. Whilst a book number is unique, there are many chapters and verses with the same number. If B represents the book number, C the

---

[5] S. E. Jones. *The Biblical Meaning of Numbers from One to Forty*, God's Kingdom Ministries, Minnesota, USA, 2008.

chapter number and V the verse number, then it is actually the set $L := \{ B, C, V \}$ that uniquely locates the verse V.

Definition 1.4 *The set $L := \{ B, C, V \}$ is the unique location in the Bible of the verse V of chapter C of the book B.*

Dealing with three numbers at once can be difficult. So, we will represent the set L by $I := B + C + V$ and call it an *identifier* of the verse. The total gematria of a verse is also an identifier of the verse.

Definition 1.5 (Method of Verse Identification) *Let Q be a set of verses in the Bible. An identifier I of Q is either (i) the total of the sum of book numbers, the sum of chapter numbers and the sum of verse numbers of all the verses in the set Q, or (ii) the total gematria (Hebrew and/or Greek) of the verses in Q.*

Example 1.1 *Consider the set Q = {Isaiah 1:19, James 2:17, 1 Kings 17:4}. The sums 153 and 9,810 are identifiers of the set Q* (see Table 1.1 below, and Tables 1.2 – 1. 4 in the Appendix for the gematriot of the original texts in Hebrew, Greek and Hebrew, respectively).

Table 1.1: Identifiers 153 and 9,810 of a set of verses

	Book B	Chapter C	Verse V	B+C+V	Total gematria
Isa 1:19	23	1	19	*43*	2076
Jam 2:17	59	2	17	*78*	4730
1Ki 17:4	11	17	4	*32*	3004
TOTAL	93	*20*	*40*	**153**	**9,810**

The MVI is a general method for it also incorporates the method of gematria given in option (ii) which recognizes the importance of each word in a verse. The advantage of option (i) is that it is independent of bible versions and translations, and independent of the original biblical text, whether Hebrew (Old Testament) or Greek (New Testament). The sum, being an identifier, simply represents a core message of the verse or set of verses, without the need to resort to gematria to find a numerical identifier of that message.

It is clear that an identifier does not identify only one set of verses. It can also be the identifier of other verses. Whilst this gives flexibility to link concepts and ideas between different sets of verses within a particular context or theme, the danger is that one may end up with completely nonsensical results. The framework of biblical mathematics shown in Fig.1.1 is therefore of paramount importance in filtering out nonsense from acceptable results. For instance, assume that after an application of the MVI (Definition 1.5), one arrives at the conclusion that God contradicts Himself. This result falls outside of the framework because by Axiom 1.1, God *is* love.

Method of the Prime

Often we want to see whether there are other verses that are linked contextually or thematically (literal or figurative) with the verses in set Q. To this end, we develop a new method inspired by a remarkable mathematical relationship between the very first verse of the Bible, namely Genesis 1:1 (KJV),

> ^{1} In the beginning God created the heaven and the earth.

and John 1:1,

> ^{1} In the beginning was the Word, and the Word was with God, and the Word was God.

The importance of these two verses cannot be emphasized enough because in John 1:14, we learn that the Word – Jesus Christ – then became flesh:

> ^{14} And the Word was made flesh, and dwelt among us, (and we beheld his glory, the glory as of the only begotten of the Father,) full of grace and truth.

Our sole axiom (Axiom 1.1), the foundation of biblical mathematics, reflects John 1:14 which is a cornerstone of Christianity. Therefore, any mathematical relationship existing between Genesis 1:1 and John 1:1 is of fundamental importance in biblical mathematics. Via the Method of the

Gematria, we outline one such relationship, given in Example 1.2 below, beginning with the following definition:

Definition 1.6 Let *Prime(N) denote the Nth prime number.*

Example 1.2 In Genesis 1:1, the well-known Hebrew word for "God" is אלהים, *Elohim*. Its gematria is $N_1:=86$. In the original Greek text for John 1:1, the well-known Greek expression for "the Word" is ο λογος, *ho logos*. Its gematria is $N_2:=70+373=443$. Then we have the unexpected result that

$$Prime(N_1) = N_2 \text{ or } Prime(86) = 443$$
Equation (1)

"God" and the "the Word" are directly related and this relationship is represented by equation (1) which is a mathematical operation that says that *Jesus is God*. This operation, in words, is as follows:

> The Hebrew gematria of "God" is the ordinal number (86) of the prime number (443) that is the Greek gematria of "the Word".

This mathematical operation is the basis of the Method of the Prime.

Definition 1.7 (Method of the Prime) *Let V_1 and V_2 be two verses in the Bible. Let N_1 be the total gematria of a set of words in V_1. Let N_2 be the total gematria of a set of words in*

V_2. It is possible that V_1 and V_2 are contextually or thematically linked if and only if $Prime(N_1) = N_2$.

From the above methodology, we propose a powerful theorem that links any quantity with an identifier of a verse, defined precisely as follows.

Theorem 1.8 (Transcendental Method of the Prime) *Let N be a quantity, with N being a natural number. Then $Prime(N)$ can be an identifier of a set of verses in the Bible. Conversely, let $Prime(N)$ be an identifier of a set of verses in the Bible. Then N can be an identifier of a set of verses in the Bible.*

Proof There are a finite number of verses in the Bible. Let T be either the total of the sum of all the book numbers, the sum of all the chapter numbers and the sum of all verse numbers, or the total gematria of all the verses. Thus $Prime(N)$ can either be smaller than or equal to T, in which case it can be an identifier, or larger than T, in which case it is not an identifier. On the other hand if $Prime(N)$ is an identifier, then it is clear that $Prime(N)$ is less than or equal to T and N is less than T. Therefore, N can be an identifier. *QED*

Theorem 1.8 is mathematically developed from the numerical finding that *Jesus is God* (Example 1.2). It is transcendental in the sense that it transcends the Basic Axioms of Algebra and units attached to a quantity (e.g., N_1=86 "Hebrew gematria", N_2=443 "Greek gematria") and therefore can capture the figurative language of the Bible and abstract concepts therein.[6] It uses ordinal numbers and their associated prime numbers to render a quantity dimensionless. In the physical sciences, this approach is not possible.[7]

In the book, we will be using the terminology "Method of the Prime" to refer to either Definition 1.7 or Theorem 1.8. The context clarifies the exact method being applied.

Summary of Results

The 10th book of the New Testament is called the Epistle to the Ephesians, the authorship of which is traditionally attributed to Paul the Apostle. It contains a series of letters from Paul to the Church of Ephesus, focusing on the necessity of, and the reasons for, the members to be pure and holy and maintain unity within the Church which he considered to be the Body of Christ. Verses 11 – 13 of chapter 2 summarize well the message of Paul (Ephesians 2: 11 – 13, New KJV):

> ^{11} Therefore remember that you, once Gentiles in the flesh—who are called Uncircumcision by what is

---

[6] Transcendental numbers and functions are part and parcel of mathematics.
[7] Mathematics is not constrained by physical impossibilities and laws. This is a fact. For example, the Axiom of Choice, in the branch of mathematics called Set Theory, roughly says that given a set of distinct infinite number of nonempty sets, we can always form a set of exactly one element from each distinct set. But it is physically impossible to go through each set to pick an element because we have an infinite number of sets!

called the Circumcision made in the flesh by hands—
¹² that at that time you were without Christ, being aliens from the commonwealth of Israel and strangers from the covenants of promise, having no hope and without God in the world. ¹³ But now in Christ Jesus you who once were far off have been brought near by the blood of Christ.

However, it is Ephesians 2: 8 – 9 that we focus on in this book:

⁸ For by grace you have been saved through faith, and that not of yourselves; *it is* the gift of God, ⁹ not of works, lest anyone should boast.

The authors consider these verses as two of the most important verses in the Bible, for they provide the precise formula for spiritual salvation: *we are saved by God's gift of grace which is a function of faith.*

The verses naturally throw up two questions:

1. What is precisely the *faith* referred to in Ephesians 2:8?
2. Given that one's deed (action or performance, as opposed to words) is not a prerequisite for God's gift of salvation grace, *how* then does one *say* or *proclaim* this faith and *to whom* should one proclaim this faith?

The objective of this book is to show that an outcome of biblical mathematics is the astonishing conclusion that the Lord's Prayer, authored by no sinful mortal but by Jesus Christ and therefore by God himself, is *the foremost* proclamation of faith that is referred to in Ephesians 2:8. We show that in praying the Lord's Prayer, we are in fact proclaiming our faith or belief that the Son fulfilled the will of the Father. Thus, the Lord's Prayer encompasses the teachings and exhortations of Paul the Apostle (Romans 10: 8 – 10, New KJV):

> [8] But what does it say? "The word is near you, in your mouth and in your heart" (that is, the word of faith which we preach): [9] that if you confess with your mouth the Lord Jesus and believe in your heart that God has raised Him from the dead, you will be saved. [10] For with the heart one believes unto righteousness, and with the mouth confession is made unto salvation.

We end by showing that indeed the Lord's Prayer is a *covenant* that Jesus Christ made with us. The Lord's Prayer completes the formula for salvation in Ephesians 2: 8 – 9.

In another astonishing outcome of biblical mathematics, we show that there is a daily time-dependent prayer pattern, given by the set of times {10.35am, 10.53am, 1.35pm, 1.53pm, 3.15pm, 3.51pm, 5.13pm, 5.31pm}. We show that the pattern of 8 explicit times represents practically the last 7 hours of the agony of Jesus Christ on the cross, noting that there are 6 hours 56 minutes from 10.35am to 5.31pm. It is shown, for the first time, that Jesus Christ died at precisely at 3.15pm[8] and His

---

[8] A 1992 study led by a Cambridge University physicist, Professor Colin Humphreys, based on astronomical calculations, concluded that Christ died at the exact time when Passover lambs were slain between 3pm and 5pm on the 14th day of the Jewish month of Nisan. [C. Humphreys and W. G.

heart was pierced at 5.31pm. The pattern is remarkably a superset of the set of permutations of the digits of the numeric 153 that appears in John 21:11, the verse that narrates a seemingly trivial final act of Jesus after His resurrection and just before His ascension.

Waddington, *The Jewish calendar, a lunar eclipse and the date of Christ's crucifixion*, Tyndale Bulletin 43 2 (1992) 331-351.]

CHAPTER 2: The Number 153 and the Will of the Father

We begin this chapter with an intriguing account of 153 fishes in the Gospel of John.

Jesus had appeared for the third time to His disciplines after His resurrection. They had been out fishing in the Sea of Galilee the whole night until morning but with no catch. The story is told in John 21, beginning with Simon Peter urging the others to go fishing with him (verses underlined by the authors for emphasis and/or future reference):

> ³ Simon Peter saith unto them, I go a fishing. They say unto him, We also go with thee. They went forth, and entered into a ship immediately; and that night they caught nothing. ⁴ But when the morning was now come, Jesus stood on the shore: but the disciples knew not that it was Jesus. ⁵ <u>Then Jesus saith unto them, Children, have ye any meat</u>? They answered him, No. ⁶ And he said unto them, Cast the net on the right side of the ship, and ye shall find. They cast therefore, and now they were not able to draw it for the <u>multitude of fishes</u>.

The author writes about the *multitude* of fishes and goes on to provide an explicit amount caught:

> ⁸ And the other disciples came in a little ship; (for they were not far from land, but as it were two hundred cubits,) dragging the net with fishes. ⁹ As soon then as they were come to land, they saw a fire of coals there,

> and fish laid thereon, and bread. ¹⁰ Jesus saith unto them, Bring of the fish which ye have now caught. ¹¹ Simon Peter went up, and drew the net to land full of great fishes, <u>an hundred and fifty and three</u>: and for all there were so many, yet was not the net broken.

Why was it necessary to include the exact number of fishes caught? The information is redundant and the number seems arbitrary.

After Simon Peter pulled in the 153 fishes, Jesus invited them for breakfast:

> ¹² Jesus saith unto them, Come and dine. And none of the disciples durst ask him, Who art thou? knowing that it was the Lord. ¹³ Jesus then cometh, and taketh bread, and giveth them, and fish likewise.

Going back to John 21:5, Jesus had adopted a word not commonly used nowadays to describe fish or food in general – *meat*:

> ⁵ Then Jesus saith unto them, Children, have ye any <u>meat</u>?

Jesus had said the same word before when, after a hectic day, His disciples asked Him whether He had eaten, as recorded in John 4:31 – 34 (KJV):

> ³¹ In the mean while his disciples prayed him, saying, Master, eat. ³² But he said unto them, I have <u>meat</u> to eat that ye know not of. ³³ Therefore said the disciples one

to another, Hath any man brought him ought to eat? [34] Jesus saith unto them, My meat is to do the will of him that sent me, and to finish his work.

The work given by the Father is described in John 6, specifically in the following verses (International Standard Version, ISV). Jesus said:

> [38] I have come down from heaven, not to do my own will, but the will of the one who sent me. [39] And this is the will of the one who sent me that I should not lose anything that he has given me, but should raise it to life on the last day. [40] This is my Father's will: That everyone who sees the Son and believes in him should have eternal life, and I will raise him to life on the last day."

In verse 44, Jesus describes *how* the Father chooses those to give to Jesus – via the Father's sovereignty in salvation:

> [44] No one can come to me unless the Father who sent me draws him, and I will raise him to life on the last day.

In John 17:1 – 2, we read that the Father gave the authority to the Son to judge all those the Father gave the Son:

> [1] After Jesus had said this, he looked up to heaven and said, "Father, the hour has come. Glorify your Son, so that the Son may glorify you. [2] For you have given him authority over all humanity so that he might give eternal life to all those you gave him.

The phrase "hour has come" in the first verse above is a reminder of Jesus' acceptance of the will of the Father for the Him to be crucified for our sins so that we may have eternal life. Jesus had the option to reject the will of the Father, for it is a *dispositional* or *preferred* will of the Father.

The above verses reveal a two-step process that leads to eternal life via the Father's will for His Son. Firstly, the Father exercises His sovereignty to choose those He wants to have eternal life through His Son, and secondly, it is through the free will of those chosen to believe in His Son to have eternal life. So *both* God's sovereignty in salvation and the free will of an individual determine whether the person will achieve eternal life.

The definition of the *will* of the Father for His son can therefore be precisely defined as follows:

Definition 2.1 (The Will of the Father for the Son) *The will of the Father for His Son, Jesus Christ, comprises the following:*

1. *The Father, by His sovereignty in salvation, chooses those He gives the Son. The will of the Father is that the Son should not lose them but to raise them to life on the last day by exercising His authority given to Him by the Father* (John 6:39, John 6:44 and John 17:2);
2. *Those chosen by the Father and given to the Son are aware of the Son. The will of the Father is that those who see and believe in the Son should have eternal life*

and be raised to life on the last day (John 6:40 and Matthew 26:39).

So when Jesus shared some of the 153 fishes that morning with His disciples, the action makes it a metaphor for the fulfillment of the will Father in Jesus Christ who had said "My meat is to do the will of him that sent me, and to finish his work." (John 4:34). Jesus had died for our sins and His Father had raised Him from the dead. Jesus had completed His earthly work as our High Priest by offering up His body as per the will of His Father. And the impact of that work is everlasting life for the believers.

That morning, when the disciples shared the fishes with the resurrected Jesus, they were witnesses to an event that is in fact the most critical of all events in the Bible because it marks the precise moment humankind can be justified before God, an event that is the culmination of a prolonged process of reconciliation between humankind and God after the fall of Adam and Eve (Romans 6:25, NIV):

> [25] He was delivered over to death for our sins and was raised to life for our justification.

The event has to be eternally remembered by humankind, and in the absence of a precise date, what a better way to do this than to seal the event with a numeral! The explicit number of fishes, 153, is the commemoration of this momentous event! The number 153 is the very signature of the event, giving us the chance to fully explore the impact of the event on humankind from a mathematical perspective.

In John 6:39, we read,

> ^{39} And this is the will of the one who sent me that I should not lose anything that he has given me, but should raise it to life on the last day.

The statement *I shall not lose anything* is an enumeration statement. Every individual given to the Son by the Father should be accounted for. So Jesus had commanded His disciples to cast the over the net to the side of the boat, for it was where the fishes had gathered, and to haul in the net with the 153 fishes towards him standing on the shore. And the net did not break, for Jesus could not afford to lose any fish if He were to fulfil the will of the Father. We have therefore a scenario which is a clear allegory of the Father's will for His Son. All those chosen by the Father are gathered up and delivered to His Son to be raised on the last day and have everlasting life. The resurrection of the Son is the decisive proof of the fulfillment of the will of the Father in His Son. By John 4:34 – *My meat is to do the will of him that sent me, and to finish his work* – the number of fishes, *Jesus' meat*, therefore precisely represents that fulfillment of the will of the Father in His Son.

Thus, from an outwardly so trivial an action of Jesus Christ – His last meal before He ascended to heaven in His 6th and final appearance to His disciples – we reached a startling conclusion: the seemingly arbitrary number 153 is not arbitrary after all; it represents the very heart of the Christian faith!

Theorem 2.2 *The number 153 in John 21:11 represents the fulfillment of the will of the Father in His Son, Jesus Christ.*

Through the fulfillment of His will in His Son, the Father declared us righteous in His sight. That is, the death and resurrection of Jesus Christ is our *justification*, the declaration of God that we are free of guilt and penalty of sin and acceptable to Him (Romans 6:25, NIV):

> [25] He was delivered over to death for our sins and was raised to life for our justification.

Our justification also marks the beginning of sanctification, a continual process of being made holy by the power of the Holy Spirit, a lifelong process that makes us more and more like Jesus Christ. In the next chapter, we prove that the Lord's Prayer is the foremost declaration of our faith in the fulfilment of the will of the Father in His Son. Therefore, it is a means to justify ourselves to the Father. We shall also prove that via the Lord's Prayer, prayed daily, we are led by the Holy Spirit in our daily lives and become more like Jesus Christ – the precise outcome of sanctification!

CHAPTER 3: The Lord's Prayer and the Will of the Father

The Lord's Prayer can be found in Matthew 6:9 – 13 and Luke 11: 2 – 4. Let us first analyze the former (KJV):

> [9] After this manner therefore pray ye: Our Father which art in heaven, Hallowed be thy name. [10] Thy kingdom come, Thy will be done in earth, as it is in heaven. [11] Give us this day our daily bread. [12] And forgive us our debts, as we forgive our debtors. [13] And lead us not into temptation, but deliver us from evil: For thine is the kingdom, and the power, and the glory, for ever. Amen.

To find the identifier of the Lord's Prayer in Matthew 6:9 – 13, we refer to Table 3.1:

Table 3.1: The identifier of the Lord's Prayer in Matthew 6:9 – 13 is 285

Book	Chapter	Verse	Sum
40	6	9	*55*
40	6	10	*56*
40	6	11	*57*
40	6	12	*58*
40	6	13	*59*
		Total	**285**

The number 285 is the identifier of the Lord's Prayer in Mathew 6:9 – 13. In base-7, the 285th number is 555. That is,

$$555_7 = 5 \times 7^2 + 5 \times 7^1 + 5 \times 7^0 = 285$$

(Equation 2)

We see here how 5 and 7 are combined mathematically into one number 285. As biblical numbers, the number 5 means "grace", and the number 7 is a symbol of "spiritual perfection". Ephesians 2:8-9 is clear: *grace comes from God.* And the Holy Spirit is the foremost embodiment of spiritual perfection, as implied by Matthew 5:48:

> [48] Therefore you shall be perfect, just as your Father in heaven is perfect.

The identifier of this verse is 93=40+5+48. The number 93 surprisingly is the gematria of the word "thelēma" (θέλημά = 93), the Greek for "will" of the Father in Matthew 6:10:

> [10] Thy kingdom come, Thy will be done in earth, as it is in heaven.

Does this connection of ideas imply that the will of the Father is perfect? Surely it is, for if the Father is perfect, then His will must be perfect as well. Thus, we have our first result in this chapter:

Theorem 3.1 *The will of the Father is perfect.*

Proof By Axiom 1.1, the will of the Father must be perfect for otherwise Jesus would not have come in the flesh. To see this, we note that the Greek for the phrase "Thy will" in the Lord's Prayer (Matthew 6:10) is made up of three words. They are τὸ θέλημά σου, *to thelēma sou*, literally "the will of-you". The total Greek gematria of these three words is 1133=370+93+670. The ordinal of 1133 is the prime number 8951. By the Method of the Prime, surprisingly we find that there is exactly one Bible verse with total gematria 8951. It is Mark 14:49 (see Table 3.2 in the Appendix), referring to Jesus' betrayal and arrest in Gethsemane, given below with verse 48 for contextual purpose:

> [48] Then Jesus answered and said to them, "Have you come out, as against a robber, with swords and clubs to take Me? [49] I was daily with you in the temple teaching, and you did not seize Me. But the Scriptures must be fulfilled."

Jesus came in the flesh to fulfil the Scriptures, just as the Father had willed for His Son. God is perfect! *QED*

Theorem 3.1 is supported biblically by Romans 12:2 (New KJV):

> [2] And do not be conformed to this world, but be transformed by the renewing of your mind, that you may prove what *is* that good and acceptable and perfect will of God.

Going back to equation (2), we see that the number 7 is raised to non-negative powers 0, 1 and 2. So 7^0, 7^1 and 7^2 are powers of 7. Hence each of the terms 5×7^2, 5×7^1 and 5×7^0 literally means that "grace is multiplied by some power of the Holy Spirit". Now,

$$555_7 = 5 \times (7^2 + 7^1 + 7^0) = 5 \times 57 = 285$$

This shows that grace is multiplied by a sum or a total of the powers of the Holy Spirit, or simply, a total power of the Holy Spirit. This prompts us to consider the following sum:

$$P := (7^{n-1} + 7^{n-2} + 7^{n-3} + \ldots + 7^0), n \text{ is a non-negative integer}$$
(Equation 3)

The sum P gives us an indication of the total power of 7. In some cases, a term can be missing from it. For instance,

$$2020_7 = 2 \times (7^3 + 7^1)$$

Regardless, these cases, including equation (3), are still partial expressions of the ultimate expression with infinite terms when we take n to infinity, an expression which should suitably be called the *infinite power of the Holy Spirit*, for it captures perfectly the power of the Almighty (Revelation 1:8):

> [8] "I am the Alpha and the Omega, *the* Beginning and *the* End," says the Lord, "who is and who was and who is to come, the Almighty."

So we will define expressions of the form of (3) or parts thereof as simply *the power of the Holy Spirit*.

Definition 3.2 *An expression of the form of equation (3) or that containing any single term or summations of some terms in (3), represents the power of the Holy Spirit.*

Equation (3) is the mathematical definition of the *power* of 7 (the Holy Spirit). But does the Bible have a precise definition of *the power* of *the Holy Spirit?* In John 16:12 – 15, we have a clear definition: it is the *authority of Jesus Christ* (New KJV):

> [12] "I still have many things to say to you, but you cannot bear *them* now. [13] However, when He, the Spirit of truth, has come, He will guide you into all truth; for He will not speak on His own *authority,* but whatever He hears He will speak; and He will tell you things to come. [14] He will glorify Me, for He will take of what is Mine and declare *it* to you. [15] All things that the Father has are Mine. Therefore I said that He will take of Mine and declare *it* to you.

Definition 3.3 *The power of the Holy Spirit is the authority of Jesus Christ.*

Thus, equation (2) means the following:

A product of God's grace and the power of the Holy Spirit is an identifier, I, of the Lord's Prayer. Precisely,

$$555_7 = 5 \times (7^2 + 7^1 + 7^0) = I$$

Thus we have before us an incredible result: in equation (3), that is, in the Lord's Prayer (555_7) we see that the power of the Holy Spirit multiplies or increases the faith $[(7^2 + 7^1 + 7^0) \times 5]$ from the Father. So equation (3), in effect, is the proof of our next substantial result in this section:

Theorem 3.4 *When we pray the Lord's Prayer, the grace we receive from the Father is multiplied or increased by the power of the Holy Spirit.*

When we apply Definition 3.3, then we see the full glory and the extraordinary power of the Lord's Prayer: *the Father's grace we receive is multiplied or increased by the authority of Jesus Christ. Through the Son, we receive an abundance of grace from the Father.*

There is indeed a verse in the Bible that refers to the multiplication of grace. It is found in the Second Epistle of Peter, in verse 2 of chapter 1 (2 Peter 1:2):

> [2] Grace and peace be multiplied unto you through the knowledge of God, and of Jesus our Lord,

In opening the Epistle with these words, the author Apostle Peter knew that it was the Holy Spirit who imparted the knowledge of God and of Jesus Christ to believers (see John 16: 12 – 15 above). With this understanding and with Theorem 3.4, we have thus the first beautiful synthesis of mathematics and the scripture:

Corollary 3.5 *When we pray the Lord's Prayer, the grace and peace we receive from the Father is multiplied or increased by the power of the Holy Spirit.*

So when we pray the Lord's Prayer, the grace and peace of the Father are ours in abundance through the power of the Holy

Spirit who imparts the knowledge of the Father and of Jesus our Lord to us.

There are several other implications of Theorem 3.4 or related results. We consider five of them under the following subsections:

1. **Sovereignty of the Father**;
    - **Corollary 3.6** *When we pray the Lord's Prayer, we acknowledge the sovereignty of the Father.*
2. **Fulfillment of the Will of the Father**;
    - **Corollary 3.7** *When we pray the Lord's Prayer, we proclaim our faith in the fulfillment of the will of His Father in His Son, Jesus Christ.*
    - **Corollary 3.8** *The Lord's Prayer is the foremost proclamation of faith in the fulfillment of the will of His Father in His Son, Jesus Christ.*
3. **Acknowledgment of the Father**;
    - **Corollary 3.9** *When we pray the Lord's Prayer, we acknowledge the Father and His love for us.*
4. **Becoming more like Jesus Christ**;
    - **Corollary 3.10** *When we pray the Lord's Prayer, we are led by the Holy Spirit to become more like Jesus Christ.*
5. **Fullness of Jesus Christ**;
    - **Theorem 3.11** *When we pray the Lord's Prayer, the grace we receive out of the fullness of Jesus Christ is multiplied or increased by the power of the Holy Spirit.*

Sovereignty of the Father

The existence of God's sovereignty is implied by Axiom 1.1. In this section, we show the Lord Prayer is an acknowledgment of this sovereignty.

Corollary 3.6 *When we pray the Lord's Prayer, we acknowledge the sovereignty of the Father.*

Proof The identifier of the Lord's Prayer, 285, is the ordinal of the prime number 1867. By the Method of the Prime, the number 1867 is also the total Greek gematria of the verse Romans 9:18 (see Table 3.3 in the Appendix):

> [18] Therefore God has mercy on whom he wants to have mercy, and he hardens whom he wants to harden.

But this is exactly a display of God's sovereignty. *QED*

Studying the Lord's Prayer carefully, we realize that it is indeed a compact literal acknowledgment of the Father's sovereignty; we refer to His Kingdom, we refer to His will, we refer to His power to sustain our lives, we refer to His power to forgive our sins, the same power that will enable us to forgive others, we refer to His power to intervene in our lives in order to straighten our path, away from temptation, and His power to snatch us away from evil. His sovereignty impacts our lives daily because He is in complete control of them.

Fulfillment of the Will of the Father

The implication of Theorem 3.4 is clear. When we pray the Lord's Prayer, we receive the grace of the Father in abundance. Therefore it is a proclamation of faith through which one gets God's gift of grace promised in Ephesians 2: 8 – 9 in order to be saved:

> ^{8} For by grace you have been <u>saved</u> through faith, and that not of yourselves; *it is* the gift of God, ^{9} not of works, lest anyone should boast.

Corollary 3.7 *When we pray the Lord's Prayer, we proclaim our faith in the fulfillment of the will of His Father in His Son, Jesus Christ.*

Because the faith – that Jesus fulfilled the will of the Father and reconciled us to the Father – is *the* foundation of Christianity, the Lord's Prayer has to be the *foremost* proclamation of faith.

Corollary 3.8 *The Lord's Prayer is the foremost proclamation of faith in the fulfillment of the will of His Father in His Son, Jesus Christ.*

When we pray the Lord's Prayer, we proclaim that Jesus suffered and died for our sins, rose from the dead, ascended to heaven, and will come again to judge the living and the dead, as per the will of the Father for the Son. The Lord's Prayer, therefore, encompasses the teachings and exhortations of Apostle Paul in the 10th Chapter of the Epistles to the Romans (Romans 10:5 – 13, NIV):

> [5] Moses writes this about the righteousness that is by the law: "The person who does these things will live by them." [6] But the righteousness that is by faith says: "Do not say in your heart, 'Who will ascend into heaven?'" (that is, to bring Christ down) [7] "or 'Who will descend into the deep?'"(that is, to bring Christ up from the dead). [8] But what does it say? "<u>The word is near you; it is in your mouth and in your heart,</u>" <u>that is, the message concerning faith that we proclaim:</u> [9] <u>If you declare with your mouth, "Jesus is Lord," and believe in your heart that God raised him from the dead, you will be saved.</u> [10] <u>For it is with your heart that you believe and are justified, and it is with your mouth that you profess your faith and are saved.</u> [11] As Scripture says, "Anyone who believes in him will never be put to shame." [12] For there is no difference between Jew and Gentile—the same Lord is Lord of all and richly blesses all who call on him, [13] for, "Everyone who calls on the name of the Lord will be saved."

Consequently, the Lord's Prayer is, firstly, a *creed* – it is formal statement of one's belief or faith, and secondly, it is a *mathematician's creed* because this faith is based on accepting Corollaries 3.7 and 3.8, which are outcomes of biblical mathematics.

Corollary 3.8 and Theorem 2.2[9] (reproduced as a footnote for convenient referencing) link the Lord's Prayer with the number 153. Precisely, *the Lord's Prayer proclaims our faith in what the number 153 represents.*

Acknowledgment of the Father

We shall show that the Lord's Prayer is *how* we acknowledge God with the very first four lines constituting the succinct acknowledgment of God (Luke 11:2):

Our Father which art in heaven.
Hallowed be thy name.
Thy kingdom come.
Thy will be done, as in heaven, so in earth.

Table 3.4 (see Appendix) provides the gematriot of the Greek words in Luke 11:2. The shaded area in the table contains the words in the above lines, giving the total gematria of 13,326. Astonishingly, we have one and only one verse in the Bible whose total gematria is 13,326 that points directly to the suffering, death and resurrection of Jesus Christ! It is Mark 10:34 (see Table 3.5, Appendix):

> [34] And they shall mock him, and shall scourge him, and shall spit upon him, and shall kill him: and the third day he shall rise again.

---

[9] Theorem 2.2 *The number 153 in John 21:11 represents the fulfillment of the will of the Father in His Son, Jesus Christ.*

By Corollaries 3.7 and 3.8, when we pray the Lord's Prayer, we acknowledge that Jesus suffered and died for our sins, ascended to heaven and will come again to judge the living and the dead. The Lord's Prayer is not only our acknowledgement of the Father, but indeed, it is also our acknowledgment of His love for us as per Axiom 1.1 and therefore John 3:16.

Corollary 3.9 (Acknowledgement of the Father) *When we pray the Lord's Prayer, we acknowledge the Father and His love for us.*

We are left to our own devices when we turn away from the Father and His love (Romans 1: 28, English Standard Version, ESV):

> [28] And since they did not see fit to acknowledge God, God gave them up to a debased mind to do what ought not to be done.

In fact, it is astonishing that the Method of Verse Identification yields the most loathsome of biblical numbers, namely 666, as an identifier of all things that God hates. These are listed in the verses 24 to 32 of Romans 1 (ESV):

> [24] Therefore God gave them up in the lusts of their hearts to impurity, to the dishonoring of their bodies among themselves, [25] because they exchanged the truth about God for a lie and worshiped and served the creature rather than the Creator, who is blessed forever! Amen. [26] For this reason God gave them up to

dishonorable passions. For their women exchanged natural relations for those that are contrary to nature; [27] and the men likewise gave up natural relations with women and were consumed with passion for one another, men committing shameless acts with men and receiving in themselves the due penalty for their error. [28] <u>And since they did not see fit to acknowledge God, God gave them up to a debased mind to do what ought not to be done.</u> [29] They were filled with all manner of unrighteousness, evil, covetousness, malice. They are full of envy, murder, strife, deceit, maliciousness. They are gossips, [30] slanderers, haters of God, insolent, haughty, boastful, inventors of evil, disobedient to parents, [31] foolish, faithless, heartless, ruthless. [32] Though they know God's righteous decree that those who practice such things deserve to die, they not only do them but give approval to those who practice them.

Table 3.6 provides the identifier of Romans 1: 24 – 32:

Table 3.6 The number 666 is the identifier of Romans 1: 24 – 32

Book	Chapter	Verse	Sum
45	1	24	*70*
45	1	25	*71*
45	1	26	*72*
45	1	27	*73*
45	1	28	*74*
45	1	29	*75*
45	1	30	*76*
45	1	31	*77*
45	1	32	*78*
		TOTAL	**666**

Does the identifier 666 identify those who do those things listed in Romans 1: 24 – 32? The answer is obvious. Corollary 3.9 provides a means to return to God and turn our back from every evil deed listed in the verses.

Becoming More Like Jesus Christ

Let us return to Ephesians 2: 8 – 9:

> [8] For by grace you have been saved through <u>faith</u>, and that not of yourselves; *it is* the gift of God, [9] not of works, lest anyone should boast.

The word "faith" in Greek is πίστεως·, *pisteōs*. The gematria of πίστεως is 1595. The 1595th prime number is 13,457. Via the Method of the Prime, we see that surprisingly there is one and only one verse in the Bible with total gematria 13,457. It is 2 Corinthians 3:18 (New JKV):

> [18] But we all, with unveiled face, beholding as in a mirror the glory of the Lord, are being transformed into the same image from glory to glory, just as [l]by the Spirit of the Lord.

The gematriot are given in Table 3.7 shown in the Appendix.

At first reading, it appears that there is no link between "faith" and the main point of the above verse, namely, "we are transformed to the same image from glory to glory." A deeper analysis though revealed an intriguing result that links "faith" with "freedom", and "freedom" with "being transformed." We have just seen that "faith" in Ephesians 2:8 is precisely the belief that Jesus fulfilled the will of the Father, as stated in Corollaries 3.7 and 3.8. These corollaries imply that we believe in the Father and in the Son foremost and believe in what they did for our salvation as per Axiom 1.1 or John 3.16. The Bible then opens up a new world that links "faith in Jesus" or "belief in Jesus" with "freedom from sin". For example, we read in Galatians 3:22 (Easy-to-Read Version, ERV):

> ^{22} But this is not possible. The Scriptures put the whole world in prison under the <u>control of sin</u>, so that the only way for people to get what God promised would be through <u>faith in Jesus</u> Christ. It is given to those who <u>believe in him</u>.

In Acts 13: 38 – 39, we read (NIV):

> ^{38} "Therefore, my friends, I want you to know that through Jesus the forgiveness of sins is proclaimed to you. ^{39} Through him <u>everyone who believes is set free from every sin</u>, a justification you were not able to obtain under the law of Moses.

So reading through the 2 Corinthians 3, we begin to see the link between "faith" in Ephesians 2:8 and the transformation "into the same image from glory to glory" in 2 Corinthians 3:19. When this verse is read in context, we realize that it is our faith or belief in Jesus Christ that sets us free from sin, a justification not possible under the Mosaic law, and transforms us into His image and we become more and more like Him (2 Corinthians 3:13 – 18):

> ¹³ We are not like Moses, who would put a veil over his face to prevent the Israelites from seeing the end of what was passing away. ¹⁴ But their minds were made dull, for to this day the same veil remains when the old covenant is read. It has not been removed, because only in Christ is it taken away. ¹⁵ Even to this day when Moses is read, a veil covers their hearts. ¹⁶ But whenever anyone turns to the Lord, the veil is taken away. ¹⁷ Now the Lord is the Spirit, and where the Spirit of the Lord is, there is freedom. ¹⁸ And we all, who with unveiled faces contemplate the Lord's glory, are being transformed into his image with ever-increasing glory, which comes from the Lord, who is the Spirit.

In Greek, the word "glory" in verse 18 is δόξαν, *doxan*, which means, in this context, "God's infinite, intrinsic worth (substance or essence)" according to Strong's Concordance. The Lord's glory is the Lord's divine quality.

Paul the Apostle, in Galatians 5, elaborates more on this concept of becoming more like Jesus Christ and taking on His divine quality. And that is, if we let the Spirit of the Lord lead our lives, we become more like Jesus Christ, taking on His divine quality about which the Mosaic law has no say (Galatians 5:18):

> ¹⁸ But if you are led by the Spirit, you are not under the law.

When the Holy Spirit leads our life daily, we are "being transformed into Jesus' image with ever-increasing glory."

In summary, when we pray the Lord's Prayer, we proclaim our faith in the Father and in the Son and in what they did for our justification. We receive what the Father promises us – grace and peace which are ours in abundance through the knowledge of the Father and of the Son imparted to us by the power of the Holy Spirit. And where the Holy Spirit is, leading us in our daily lives, there is freedom because we are becoming more like Jesus whose divine quality is not covered or penalizable by any Mosaic law.

Corollary 3.10 *When we pray the Lord's Prayer, we are led by the Holy Spirit to become more like Jesus Christ.*

This result establishes that praying the Lord's Prayer is an act of acquiring holiness with the help of the Holy Spirit. Praying the Lord's Prayer is *how* we can partake in sanctification.

Lest we forget, where the Holy Spirit is, there is liberty, but with liberty comes responsibility, as Paul urges in Galatian 5: 1:

> [1] Stand fast therefore in the liberty by which Christ has made us free, and do not be entangled again with a yoke of bondage.

Fullness of Jesus Christ

The Lord's Prayer in the Gospel of Luke differs from the Lord's Prayer in the Gospel of Matthew in that the last sentence, namely, *For thine is the kingdom, and the power, and the glory, for ever*, in Mathew's version is not in Luke's version. Nonetheless, in this section, we show the amazing result that the portion is not necessary to arrive at the main conclusions about the effectiveness of the Lord's Prayer as described in Theorem 3.4[10] and Corollary 3.7[11].

In Luke 11:2 – 4, we read:

> ² And he said unto them, When ye pray, say, Our Father which art in heaven, Hallowed be thy name. Thy kingdom come. Thy will be done, as in heaven, so in earth. ³ Give us day by day our daily bread. ⁴ And forgive us our sins; for we also forgive every one that is indebted to us. And lead us not into temptation; but deliver us from evil.

The Method of Verse Identification gives the identifier 168 (Table 3.8).

Table 3.8: Identifier of the Lord's Prayer in Luke 11

Book	Chapter	Verse	Sum
42	11	2	55
42	11	3	56
42	11	4	57
			168

---

[10] **Theorem 3.4** *When we pray the Lord's Prayer, the grace we receive from the Father is multiplied or increased by the power of the Holy Spirit.*
[11] **Corollary 3.7** *When we pray the Lord's Prayer, we proclaim our faith in the fulfillment of the will of His Father in His Son, Jesus Christ.*

It is astonishing that the identifiers of the first three verses in the version of the Lord's Prayer in Matthew 6:9 – 13 are equal to the identifiers of all three verses of the Lord's Prayer in Luke 11: 2 – 4 (Table 3.9).

Table 3.9: The Lord's Prayer

Matthew 6	Luke 11	*I*
[9] After this manner therefore pray ye: Our Father which art in heaven, Hallowed be thy name.	[2] And he said unto them, When ye pray, say, Our Father which art in heaven, Hallowed be thy name. Thy kingdom come. Thy will be done, as in heaven, so in earth.	55
[10] Thy kingdom come, Thy will be done in earth, as it is in heaven.	[3] Give us day by day our daily bread.	56
. [11] Give us this day our daily bread.	[4] And forgive us our sins; for we also forgive every one that is indebted to us. And lead us not into temptation; but deliver us from evil.	57

The verses of the Lord's Prayer in Matthew 6 missing from Table 3.9 are 12 and 13.

> [12] And forgive us our debts, as we forgive our debtors.
> [13] And lead us not into temptation, but deliver us from evil: For thine is the kingdom, and the power, and the glory, for ever. Amen.

However, we note that verse 12 and the first part of verse 13 are already in verse 4 of the version of the Lord's Prayer in Luke 11. We shall now show that it is not necessary to include the last part of Matthew 6:13, namely, *For thine is the kingdom, and the power, and the glory, for ever*, to arrive at the same conclusion that the Lord's Prayer is the proclamation of our faith in the fulfillment of the will of the Father in His Son, Jesus Christ. That is, Corollary 3.7[12] still holds.

We have that $168 = 330_7$:

$$330_7 = 3 \times 7^2 + 3 \times 7^1 = 168$$

(Equation 4)

The biblical meaning of the number 3 is "divine fullness". When we read John 1:14 – 17, we understand this as the "fullness of Jesus Christ", meaning that "Jesus Christ is full of grace and truth" (KJV):

> [14] And the Word was made flesh, and dwelt among us, (and we beheld his glory, the glory as of the only begotten of the Father,) full of grace and truth. [15] John bare witness of him, and cried, saying, This was he of whom I spake, He that cometh after me is preferred before me: for he was before me. [16] And of his fulness have all we received, and grace for grace. [17] For the law

---

[12] **Corollary 3.7** *When we pray the Lord's Prayer, we proclaim our faith in the fulfillment of the will of His Father in His Son, Jesus Christ.*

was given by Moses, but grace and truth came by Jesus Christ.

John 1:16 says that out of His fullness, it is grace that we receive in abundance.

Thus equation (4) refers to the multiplication of Jesus Christ's grace (out of His fullness) by the power of the Holy Spirit:

A product of the grace out of the fullness of Jesus Christ and the power of the Holy Spirit is an identifier, I, of the Lord's Prayer. Precisely, $330_7 = 3 \times (7^2 + 7^1) = 168 = I.$

Theorem 3.11 *When we pray the Lord's Prayer, the grace we receive out of the fullness of Jesus Christ is multiplied or increased by the power of the Holy Spirit.*

It is clear that Theorem 3.11 is consistent with Theorem 3.4[13]. Therefore Corollary 3.7 also holds, independent of which version of the Lord's Prayer one uses.

---

[13] **Theorem 3.4** *When we pray the Lord's Prayer, the grace we receive from the Father is multiplied or increased by the power of the Holy Spirit.*

CHAPTER 4: The Lord's Prayer and Our Sanctification

The 153 fishes were caught in the net, as narrated in John 21:11. In Chapter 2 of this book, we showed that the number 153 represents the fulfillment of the will of the Father in His Son. What does "the net" represent?

> [11] Simon Peter went up, and drew the net to land full of great fishes, an hundred and fifty and three: and for all there were so many, yet was not the net broken.

In several passages of the Bible, the word "net" is used figuratively to mean *judgment of*. In the Book of Hosea, for instance, we read of God's *judgment of* the behavior of His people in the Northern Kingdom of Israel after it had turned away from Him and served other gods (Hosea 7:11 – 12, KJV):

> [11] Ephraim also is like a silly dove without heart: they call to Egypt, they go to Assyria. [12] When they shall go, I will spread my net upon them; I will bring them down as the fowls of the heaven; I will chastise them, as their congregation hath heard. [13] Woe unto them! for they have fled from me: destruction unto them! because they have transgressed against me: though I have redeemed them, yet they have spoken lies against me.

The "net" in the passage is ישתי, the transliteration of which is *rištî*. Its root word is תשר, *resheth*, which according to Brown-

Driver-Briggs, is also a figurative expression of "judgment of". In God's view, His people had left Him, they had transgressed against Him and they had spoken lies against Him, even though He had redeemed them. As a consequence, He would chastise them and bring destruction upon them.

Is there a metaphorical dimension to the usage of the term "the net" in John 21:11? We argue that there is; in fact, we look at two possible exegeses:

1. The Lord's Prayer is a means to judge ourselves; via it, we are the *judges of* our character, daily reflecting on whether we are measuring up to the expectations of the Father, keeping at the back of our minds that it is the very word of Jesus Christ that will judge us in the last day (John 10:48):

   > [48] He who rejects Me, and does not receive My words, has that which judges him—the word that I have spoken will judge him in the last day.

   Realizing that as sinful mortals we cannot possibly measure up to the expectations of God, the Lord's Prayer offers us the opportunity to humble ourselves before God and ask for forgiveness.
2. Jesus Christ is the judge of humanity.

The Lord's Prayer

To the Jews, a prayer time is a time of self-judgement.[14]

The Hebrew for "prayer" is תפלה, *tephillah*. According to NAS Exhaustive Concordance, it comes from the verb ללפ, *palal* that means "to intervene, interpose", or as indicated by Brown-Driver-Briggs, "to arbitrate, judge, intercede." Its reflective verb is להתפלל, *lehitpalal,* "to judge oneself". Thus, a prayer time should be a time to examine ourselves carefully, critically and sincerely, knowing that we are all sinners and fall short of God's glorious standards (Romans 3:23):

> ^{23} For all have sinned, and come short of the glory of God;

Since we cannot possibly meet God's expectations, we do not deserve His blessings and favors. Therefore, we humble ourselves before Him and confess our sins in our prayers. Psalms 51 and 69 come to mind as some of King David's most significant prayers that are unparalleled in their long, sincere and careful examination of thoughts and feelings. It is only after first carefully, critically and sincerely examining ourselves that we can partake in sanctification with the help of the Holy Spirit.

It is from this Jewish perspective of what a prayer should be that we can start to fully appreciate the Lord's Prayer – why

---

[14] https://www.chabad.org/library/article_cdo/aid/682090/jewish/The-Meaning-of-Prayer.htm

we begin by exalting His name when we say *Our Father in heaven, holy be your name*, why we humble ourselves before Him when we say *forgive us our sins, as we forgive those who sin against us*, and why we then request the Father not *to bring us to the test but deliver us from evil*, which is exactly meant for our sanctification. To show that our work support this interpretation of the Lord's Prayer, let us recall two results from Chapters 2 and 3:

Theorem 2.2 *The number 153 in John 21:11 represents the fulfillment of the will of the Father in His Son, Jesus Christ.*

Corollary 3.7 *When we pray the Lord's Prayer, we proclaim our faith in the fulfillment of the will of the Father in His Son.*

These results link the Lord's Prayer with the number 153. That is, the Lord's Prayer proclaims the faith in that which the number 153 represents.

Next we recall three more results from Chapter 3:

Corollary 3.5 *When we pray the Lord's Prayer, the grace and peace we receive from the Father is multiplied or increased by the power of the Holy Spirit.*

Corollary 3.10 *When we pray the Lord's Prayer, we are led by the Holy Spirit to become more like Jesus Christ.*

Theorem 3.11 *When we pray the Lord's Prayer, the grace we receive out of the fullness of Jesus Christ is multiplied or increased by the power of the Holy Spirit.*

These, together with Corollary 3.7, show that praying the Lord's Prayer is how we can partake in sanctification.

Theorem 4.1 *The Lord's Prayer, as a means of self-judgment and partaking in sanctification, is an exegesis of "the net" in the bible verse Matthew 21:11.*

The net in Matthew 21:11 did not break. The unbroken net is the metaphor for praying the Lord's Prayer daily, continually. As we pray daily, we are being drawn closer, hauled in by the "unbroken net", to our Savior Jesus Christ, to be one with Him. We are becoming more and more like Him. As His "meat", when we are one with Him, we are "dead to sin, but alive to God in Jesus Christ" (Romans 6:11). The allegory of the 153 fishes in the net is clear: it is those who pray the Lord's Prayer daily for progressive sanctification towards glorification in Jesus Christ.

Corollary 4.2 *The 153 fishes in the net represent those who pray the Lord's Prayer continually for sanctification.*

We thus ask a critical question: how often should one pray the Lord's Prayer?

The Greek for "the net" in John 21:11 is το δικτυον, *to diktyon*. Surprisingly, the total Greek gematria is 370 + 854 = 1224, and

$$1224 = 153 \times 8$$

(Equation 5)

Whilst equations (2) and (4) are identifiers of the Lord's Prayer, the number in equation (5) represents the Lord's Prayer itself because we have shown that the net in John 11:21 symbolizes the Lord's Prayer. Thus, as a consequence of Theorem 4.1, we have:

Corollary 4.3 *The number* $1224 = 153 \times 8$ *represents the Lord's Prayer.*

The biblical meaning of the number 8 is "new beginning". The number 8 therefore introduces the time factor in equation (5).

We have thus arrived at an intriguing result. Since the number 153 represents the fulfillment of the will of the Father in His Son, and since the Lord's Prayer is the proclamation of our faith in what the number represents, equation (5) says that the Lord's Prayer must be prayed eight times daily.

Corollary 4.4 *The Lord's Prayer is to be prayed daily eight times.*

What are these times? Before we answer this question, we first consider *how* Jesus prayed. This will be discussed in the next chapter (Chapter 5).

Jesus Christ is the Judge of Humanity

In John 17:1 – 2, we read that the Father gave the Son the authority over all humanity (ISV):

> [1]After Jesus had said this, he looked up to heaven and said, "Father, the hour has come. Glorify your Son, so that the Son may glorify you. [2] For you have given him

authority over all humanity so that he might give eternal life to all those you gave him.

So when Jesus Christ commanded His disciples to throw their net to the right side of the boat and haul in the 153 fishes, the net became the metaphor for the authority of Jesus Christ to give eternal life to all those the Father gave Him, represented by the fishes made to gather on the right-hand side of the boat. Indeed, the Father has committed *all* judgment to the Son (John 5:22).

> [22] For the Father judges no one, but has committed all judgment to the Son,

He is "the one appointed by God to be *judge of* the living and the dead" (Acts 10:42, ESV):

> [42] And he commanded us to preach to the people and to testify that he is the one appointed by God to be judge of the living and the dead.

Theorem 4.5 *The authority of Jesus Christ over all humanity is an exegesis of "the net" in the bible verse Matthew 21:11.*

By Theorem 4.1, "the net" is the Lord's Prayer that will 'haul in" those who prayed the prayer to the feet of Jesus Christ. By Theorem 4.5, "the net" represents the authority of Jesus Christ over them so He might give them eternal life. Hence, we have a clear consequence of both theorems:

Corollary 4.6 *When we pray the Lord's Prayer, we acknowledge that Jesus Christ is the judge of humanity.*

CHAPTER 5: The Lord's Prayer and the Lord's Last Prayer

In Psalm 69, we read of a soul-searching prayer credited to King David. Verses 10 to 14 are provided below (NIV):

> [10] When I wept and humbled my soul with fasting,
> it became my reproach.
> [11] When I made sackcloth my clothing,
> I became a byword to them.
> [12] I am the talk of those who sit in the gate,
> and the drunkards make songs about me.
> [13] <u>But as for me, my prayer is to you, O LORD.
> At an acceptable time, O God,</u>
> in the abundance of your steadfast love answer me in
> your saving faithfulness.
> [14] Deliver me
> from sinking in the mire;
> let me be delivered from my enemies
> and from the deep waters.

In this prayer, verse 13 is revealing. King David, filled with humility, left it to God to decide *when* God would listen to him. The verse is telling us that there is a time of prayer that is acceptable or favorable to God! One cannot help but wonder, therefore, that when Jesus prayed, were they the times acceptable to His Father? To find the answer, we revisit His last moments on earth when He was on the cross.

In Luke 23: 44 – 46, we read:

⁴⁴ And it was about the sixth hour, and there was a darkness over all the earth until the ninth hour. ⁴⁵ And the sun was darkened, and the veil of the temple was rent in the midst. ⁴⁶ And when Jesus had cried with a loud voice, he said, Father, into thy hands I commend my spirit: and having said thus, he gave up the ghost.

In Matthew 27: 45 – 50, we read:

⁴⁵ Now from the sixth hour there was darkness over all the land unto the ninth hour.⁴⁶ And about the ninth hour Jesus cried with a loud voice, saying, Eli, Eli, lama sabachthani? that is to say, My God, my God, why hast thou forsaken me? ⁴⁷ Some of them that stood there, when they heard that, said, This man calleth for Elias.⁴⁸ And straightway one of them ran, and took a spunge, and filled it with vinegar, and put it on a reed, and gave him to drink.⁴⁹ The rest said, Let be, let us see whether Elias will come to save him. ⁵⁰ Jesus, when he had cried again with a loud voice, yielded up the ghost.

Mark 15 gives a similar narration:

³³ And when the sixth hour was come, there was darkness over the whole land until the ninth hour. ³⁴ And at the ninth hour Jesus cried with a loud voice, saying, Eloi, Eloi, lama sabachthani? which is, being interpreted, My God, my God, why hast thou forsaken me? ³⁵ And some of them that stood by, when they heard it, said, Behold, he calleth Elias. ³⁶ And one ran and filled a spunge full of vinegar, and put it on a reed, and gave him to drink, saying, Let alone; let us see

whether Elias will come to take him down. ³⁷ And Jesus cried with a loud voice, and gave up the ghost.

> ²⁸ After this, Jesus knowing that all things were now accomplished, that the scripture might be fulfilled, saith, I thirst. ²⁹ Now there was set a vessel full of vinegar: and they filled a spunge with vinegar, and put it upon hyssop, and put it to his mouth. ³⁰ When Jesus therefore had received the vinegar, he said, It is finished: and he bowed his head, and gave up the ghost.

The gospels of Matthew, Mark and John all agree on one point; after Jesus had drank the vinegar, He cried out one last time, and then He died. John 19:30 gives us what He said: "It is finished." The Greek corresponding to this is τετελεσται, *tetelestai*, the literal translation of which is "it has been accomplished." Jesus, in His life and death, had fulfilled biblical prophesies of the suffering Messiah of Isaiah 53 (see Appendix for proof). In Him, the messianic prophesies had been confirmed, supported or upheld. According to Strong's Concordance, the Hebrew verb נמא, *aman*, means to confirm or to support. It is the root word for "Amen."

Thus, we are certain that that the last cries of Jesus Christ on the cross make up His last prayer. We propose therefore that Jesus prayed as follows:

Theorem 5.1 (The Lord's Last Prayer)

My God, my God, why hast thou forsaken me?
Father, into thy hands I commend my spirit.
It is finished.

To show that these lines constitute the Lord's Last Prayer (independent of order of saying), we calculate to the identifier of the verses (Table 5.1):

Table 5.1: Verses containing Jesus' last prayer

Verses		Book#	Chapter#	Verse#
My God, my God, why hast thou forsaken me?	Matthew 27:46	40	27	46
Father, into thy hands I commend my spirit	Luke 23:46	42	23	46
It is finished	John 19:30	43	19	30

The identifier of the verses is 316. Now the 316th prime number is 2089. There are a total of five verses whose Hebrew or Greek gematria is 2089. Remarkably, one of them is a scripture in the Old Testament that predicted the death of Jesus Christ – Zechariah 12:11 (Table 5.2). The scripture foretold the day of the great weeping in Jerusalem, when the inhabitants of Jerusalem would grieve bitterly for someone whom they have pierced:

> [10] And I will pour out on the house of David and the inhabitants of Jerusalem a spirit of grace and supplication. They will look on me, the one they have

pierced, and they will mourn for him as one mourns for an only child, and grieve bitterly for him as one grieves for a firstborn son. <u>¹¹ On that day the weeping in Jerusalem will be as great as the weeping of Hadad Rimmon in the plain of Megiddo.</u>

Table 5:2: The gematria of Zechariah 12:11 is 2089

Transit.	Hebrew	Gematria	English
bay-yō-wm	ביום	58	In that day
ha-hū,	ההוא	17	In that
yiḡ-dal	יגדל	47	shall there be a great
ham-mis-pêḏ	המספד	189	mourning
bî-rū-šā-lim,	בירושלם	588	in Jerusalem
kə-mis-paḏ	כמספד	204	as the mourning
hă-ḏaḏ-	הדד	13	-
rim-mō-wn	רמון	296	of Hadadrimmon
bə-ḇiq-'aṯ	בבקעת	574	in the plain
mə-ḡid-dō-wn.	מגדון	103	of Megiddo
		2089	

Zechariah 12:10 is the very scripture John 19:37 refers to!

³⁷ And again another scripture saith, They shall look on him whom they pierced.

Hence, in a surprising twist, the proposed Last Prayer of Jesus Christ connects directly to the day that the Jews would mourn His death, the death of "one they have pierced". In His death –

when He uttered "It is finished" – all biblical prophecies were fulfilled.

It must be noted that via a methodology called *Gospel Harmony*, in which different Bible verses or portions of them are combined in the most likely order of occurrence, biblical scholars have identified seven sayings of the Jesus during the crucifixion and ordered them as follows:

1. *Father, forgive them, for they know not what they are doing*;
2. *Today you will be with me in paradise*;
3. *Behold your son: behold your mother*;
4. *My God, my God, why have you forsaken me?*
5. *I thirst*;
6. *It is finished*;
7. *Father, into your hands I commend my spirit.*

The method of Gospel Harmony suggests the order:

> *My God, my God, why have you forsaken me?*
> *It is finished;*
> *Father, into your hands I commend my spirit.*

The Method of Verse Identification which produces Table 5.1 does not contradict the order because it is independent of the ordering. The identifier 316 remains the same regardless of the order. Theorem 5.1 agrees with the method of Gospel harmony in the sense that the three sayings are of Jesus' before He died. We argue though that Theorem 5.1 gives the correct order because of our interpretation of the exclamation "It is finished" refers to Jesus fulfilling the scriptures (John 19:28), which in turn connects to the Hebrew word "aman", from which the word "amen" arises. Indeed, it is because of this fact that we know that Jesus actually said a *prayer* to His Father before He died.

Was the timing of prayer acceptable to His Father?

Theorem 3.1 says the he will of God is perfect in all its manifestation. Therefore the manner in which it is fulfilled must be perfect as well, for if this is not true, then the prophecies would not have been fulfilled. The fulfillment of the Scriptures of the prophets implies Jesus Christ was obedient to His Father till the very end, as we also learn from Philippians 2:8:

> [8] And being found in fashion as a man, he humbled himself, and became obedient unto death, even the death of the cross.

So when He was praying His Last Prayer on the cross, He was fulfilling His Father's will. It is logical to conclude therefore that from Theorem 3.1, Philippians 2:8 and Psalm 69:13, Jesus Christ prayed at the times acceptable to His Father, and hence, according to His Father's will. That is, it is the dispositional will of the Father that the Son should pray and do so at the times acceptable to Him.

Theorem 5.2 *It is the will of the Father that the Son should pray at the times acceptable to Him.*

Indeed, we know that whatever Jesus says not only comes from the Father but it is also a command from the Father (John 12:49 – 50, New KJV):

> [49] For I have not spoken on My own *authority;* but the Father who sent Me gave Me a command, <u>what I should say and what I should speak</u>. [50] And I know that His command is everlasting life. Therefore, whatever I speak, just as the Father has told Me, so I speak."

By extension, therefore, since Jesus Christ taught and instructed His disciples to pray the Lord's Prayer, the prayer must also be the command from the Father:

Corollary 5.3 *It is the will of the Father that we should pray the Lord's Prayer at the times acceptable to Him.*

What times are acceptable to the Father to hear us say His Son's Prayer? The answer lies in the number 153 itself!

CHAPTER 6: The Lord's Prayer and the Death of Jesus Christ

In another astonishing expression of 153, it is shown in this chapter that it generates a set of numbers that directly reflect the hours over which Jesus Christ hung in agony on the cross, the exact time of death, and the exact time when His side was pierced.

We begin by recalling two results from Chapter 4:

Corollary 4.3 *The number* $1224 = 153 \times 8$ *represents the Lord's Prayer.*

Corollary 4.4 *The Lord's Prayer is to be prayed daily eight times.*

What are these times, which according to Corollary 5.3, must be acceptable to the Father? To attempt to answer this question, we revisit the last moments of Jesus Christ on the cross. It Chapter 5, it is shown that His cries (given in Theorem 5.1) constitute His last prayer. Hence if the precise time Jesus death is known, then it would provide an estimate of the time He prayed.

We posit the idea that since the number 153 represents the fulfillment of the Father's will in Jesus Christ, the number itself can provide the answer.

Now, recall Matthew 27: 45 – 50:

> [45] Now from the sixth hour there was darkness over all the land unto the ninth hour. [46] <u>And about the ninth hour Jesus cried with a loud voice</u>, saying, Eli, Eli, lama sabachthani? that is to say, My God, my God, why hast thou forsaken me? [47] Some of them that stood there, when they heard that, said, This man calleth for Elias. [48] And straightway one of them ran, and took a spunge, and filled it with vinegar, and put it on a reed, and gave him to drink. [49] The rest said, Let be, let us see whether Elias will come to save him. [50] <u>Jesus, when he had cried again with a loud voice</u>, yielded up the ghost.

The 9th hour is accepted to be 3pm. Obviously between Jesus' first cry at 3pm and His last cry time had elapsed. Therefore, He could not have died at precisely 3pm.

It is highly likely that the action of the soldier fetching the vinegar and then getting Jesus to drink took some time, for it requires steady hands with stretched arms to hold onto the reed extended to a dying man, almost incapacitated after 6 hours on the cross, barely able to open his mouth. While Jesus struggled to drink, the soldiers mocked Him. It is not unreasonable therefore to conclude that several minutes passed before Jesus died. Given our understanding of the importance of the number 153, namely that it represents the fulfillment of the will of the Father in His Son, let us assume, for the moment, Jesus died at 3.15pm. Ignoring the decimal point for now, the digits of the numeral 315 is but a permutation of the digits of the numeral 153. But there are 6 permutations of {1, 5, 3}, namely,

$$\{1,3,5\}, \{1,5,3\}, \{3,1,5\}, \{3,5,1\}, \{5,1,3\}, \{5,3,1\}$$

The digits of the permutations, in turn, produced the numerals in the following set:

$$\{135, 153, 315, 351, 513, 531\}$$

Based on the assumed time of death at 3.15pm, the time-equivalents are therefore the elements of the following set:

$$C:= \{1.35\text{pm}, 1.53\text{pm}, 3.15\text{pm}, 3.51\text{pm}, 5.13\text{pm}, 5.31\text{pm}\}$$
(Equation 6)

From 1.35pm to 5.31pm, we have 3 hours 56 minutes, or 236 minutes. Applying the Transcendental Method of the Prime, we find that the 236th prime number is 1487, which corresponds to the Hebrew gematria of exactly three verses in the Bible (Table 6.1):

Table 6.1: Verses with gematria 1487

Text (KJV)	Verse
And Boaz begat Obed, and Obed begat Jesse,	1 Chronicles 2:12
Plead *my cause*, O LORD, with them that strive with me: fight against them that fight against me.	Psalm 35:1

> **But it shall be one day which shall be known to the LORD, not day, nor night: but it shall come to pass, *that* at evening time it shall be light.** Zechariah 14:7

It is astonishing that set *C* which contains the proposed time of Jesus' death, corresponds contextually to Zechariah 14:7 (see gematria in Table 6.2 in the Appendix) that precisely describes the day of crucifixion, when day mixed with darkness and Jesus would become the Light of the World at His death! For contextual purpose, we provide below Zechariah 14: 6 – 7 (NIV):

> ^{6} On that day there will be neither sunlight nor cold, frosty darkness. ^{7} It will be a unique day—a day known only to the LORD—with no distinction between day and night. When evening comes, there will be light.

We know that Jesus is The Light of the World from John 8:12 (NIV):

> ^{12} When Jesus spoke again to the people, he said, "I am the light of the world. Whoever follows me will never walk in darkness, but will have the light of life."

We can only arrive at one conclusion:

The proposed time of death is correct and that the set C provides a precise time interval, namely [1.35pm, 5.31pm], *over which we are sure that Jesus was on the cross. That is, set C specifically represents Jesus' ordeal on the cross, His sustained agony on the cross from after midday till before Sabbath.*

It is not unreasonable therefore to ask the following question: Could the set C be extended or generalized to represent Jesus' entire ordeal, that is, the entire time He was on the cross, given that He was crucified at 9am as per Mark 15:25? (New Living Translation, NLT):

> [25] It was nine o'clock in the morning when they crucified him.

Before 12 noon, we can clearly identify two instances of supplication addressed to the Father. In Luke 23, we read in verse 34 that Jesus addressed His Father after He was crucified:

> [33] And when they were come to the place, which is called Calvary, there they crucified him, and the malefactors, one on the right hand, and the other on the left. [34] Then said Jesus, <u>Father, forgive them; for they know not what they do</u>. And they parted his raiment, and cast lots.

When Jesus addressed His Father, was it the right time to do so? By Theorem 5.2[15], the answer is affirmative, for obviously, this is a supplication to the Father.

Jesus spoke again when one the thieves crucified alongside Him acknowledged He was the Lord (Luke 23: 42 – 43):

---

[15] **Theorem 5.2** *It is the will of the Father that the Son should pray at the times acceptable to Him.*

> [42] And he said unto Jesus, <u>Lord, remember me when thou comest into thy kingdom.</u> [43] And Jesus said unto him, Verily I say unto thee, <u>Today shalt thou be with me in paradise.</u>

The Greek for "today" is σήμερον, *sémeron,* which according to Strong's Concordance, also means "now". So when Jesus answered in the affirmative, He must have had the Father's confirmation of salvation for both of them at that instant.

These are the two recorded moments after Jesus was crucified and before 12 noon when two supplications were made to the Father – one directly from Jesus Christ and the other indirectly through Jesus Christ. The times of supplications were times acceptable to the Father.

Now, in the 12-hour format, after 9am and before 12 noon, and reminding ourselves of the number 153 and its importance, there are only two logical choices of time that can expand or generalize the set C. The choices are 10.35am and 10.53am. Indeed, taking any one of these and ignoring the decimal point for the moment, we generate the following times, by permutation of the digits of 1035 or 1053 (see Table 6.3). The times shaded are not valid with respect to the 12-hour format.

Table 6.3: Permutations of the digits of 1035 or 1053 reproduce the time-equivalent set C

10.35	10.53	13.05	13.50	15.03	15.30
01.35	01.53	03.15	03.51	05.13	05.31
31.05	31.50	30.15	30.51	35.10	35.01
51.03	51.30	50.13	50.31	53.10	53.01

Hence, we have the following generalized set whose subset is C:

$$D := \{10.35\text{am}, 10.53\text{am}, 1.35\text{pm}, 1.53\text{pm}, 3.15\text{pm}, 3.51\text{pm},\\ 5.13\text{pm}, 5.31\text{pm}\}$$

(Equation 7)

Could this set of numbers represent the ordeal of Jesus on the cross?

The time interval between 10.35am and 5.31pm is 6 hours 56 minutes or 416 minutes. The 416th prime number is 2861. Incredibly there is exactly one Bible verse that has a word with Greek gematria 2861. That word is στρατιωτῶν, *stratiōton* – soldiers – in John 19:34. In context, it refers to one of the soldiers who pierced the side of Jesus Christ to ensure that Jesus was dead:

> [34] But one of the soldiers with a spear pierced his side, and forthwith came there out blood and water.

This incredible result can only mean one thing: *that this event took place at exactly 5.31pm!* This also implies that besides the death of Jesus Christ at 3.15pm, this event must be also an important one. Reading the rest of John 19 gives us the precise reasons why the event has to happen (John 19: 31 – 37, New KJV):

> [31] Therefore, because it was the Preparation *Day,* that the bodies should not remain on the cross on the Sabbath (for that Sabbath was a high day), the Jews

asked Pilate that their legs might be broken, and *that they might be taken away.* ^{32} Then the soldiers came and broke the legs of the first and of the other who was crucified with Him. ^{33} But when they came to Jesus and saw that He was already dead, they did not break His legs. ^{34} But one of the soldiers pierced His side with a spear, and immediately blood and water came out. ^{35} And he who has seen has testified, and his testimony is true; and he knows that he is telling the truth, so that you may believe. ^{36} For these things were done that the Scripture should be fulfilled, "Not *one* of His bones shall be broken." ^{37} And again another Scripture says, "They shall look on Him whom they pierced."

The soldier was the witness. He was the third party who told the world that, yes, Jesus was real, and died on the cross. And through mathematics, we place the timing of the event at precisely 5.31pm before His body was taken down ahead of Sabbath.

The set D thus represents the *Agony of Jesus Christ on the Cross*, which includes the two critical events on cross, namely, His death and the piercing of His side.

We summarize the above discussions with the following two results, noting that 6 hours 56 minutes is within, if not practically, the last 7 hours of Jesus' life on earth.

Theorem 6.1 (Agony of Jesus Christ on the Cross) *The set of numbers D given in equation (7) represents the agony of Jesus Christ within the last 7 hours of His life on the cross.*

Theorem 6.2 *Jesus died at 3.15pm, and His side was pierced at 5.31pm.*

It is logical therefore to conclude that Jesus ended His Last Prayer on the cross at 3.15pm.

Corollary 6.3 *Jesus Christ concluded His Last Prayer on the cross at 3.15pm.*

The set D therefore gives us the answer to the question on the number of times the Lord's Prayer is to prayed, for the death of Jesus Christ and day and timing of His death are but the will of His Father.

Theorem 6.4 *The Lord's Prayer is to be prayed daily at the times given in set D.*

Since the number 8 means "new beginning", it suggests that in adopting the prayer pattern, we are renewed in the spirit daily. From this perspective of renewing our spirit daily by a new pattern of praying the Lord's Prayer, Romans 12:2 makes sense. It is given below with verse 1 for contextual purpose (New KJV):

> [1] I beseech you therefore, brethren, by the mercies of God, that you present your bodies a living sacrifice, holy, acceptable to God, *which is* your reasonable service. [2] And do not be conformed to this world, but be transformed by the renewing of your mind, that you may prove what *is* that good and acceptable and perfect will of God.

Recall that verse Romans 12:2 supports Theorem 3.1 which establishes that God's will is perfect. But it is Romans 12:1 that brings up an unexpected and astonishing conclusion. We have just ended proving that set D contains the time of the death of Jesus Christ on the cross. It is Jesus Christ who literally presents His body as "a living sacrifice, holy and pleasing to God" at the cross. So by Romans 12:1, His body, literally being offered up as living sacrifice, is the reasonable service to the Father! In a parallel argument, we have seen that Lord's Prayer is a means of sanctification, for it is via which we become more and more like Jesus Christ. So it is the Lord's Prayer, when prayed daily at the times given in set D, that presents our bodies as "a living sacrifice, holy and pleasing to God."

Theorem 6.5 *The Lord's Prayer, prayed daily at the times given in set D, presents our bodies as a living sacrifice, holy and pleasing to God.*

When we look back and consider all the results we have derived thus far, this remarkable conclusion becomes apparent. Indeed, it is unthinkable that such a very simple prayer is actually a deep reservoir of messages that are the central tenets of the Christian faith (see Table 6.4).

Table 6.4: Messages of the Lord's Prayer beyond the literal

The Lord's Prayer	Related Messages
Our Father who art in heaven. Hallowed by the name.	*We acknowledge the Father* (Corollaries 3.6 and 3.9)
Thy kingdom come.	*We receive the grace and peace of the Father and of the Son* (Corollary 3.5 and Theorem 3.11)
Thy will be done, on earth as it is in heaven.	*We acknowledge the sovereignty and the perfect will of the Father* (Corollary 3.6, Theorem 3.1). *We proclaim our faith in the fulfilment of the will of the Father in His Son Jesus Christ* (Corollaries 3.7, 3.8)
Give us this day our daily bread;	*We become more and more like Jesus Christ* (Corollary 3.10)
And forgive us our trespasses, as we forgive those who trespass against us;	*We acknowledge the authority of Jesus Christ to judge us* (Theorem 4.5).
And lead us not into temptation, but deliver us from evil.	*Our bodies are a living sacrifice, holy and pleasing to the Father* (Theorem 6.5)

CHAPTER 7: The Lord's Prayer and Prayer Times

The prayer times in set D (equation (7)) provides the following options in a 12-hour period:

1. {10.35am,10.53am,1.35pm,1.53pm,3.15pm, 3.51pm, 5.13pm, 5.31pm}; or
2. {1.35pm,1.53pm,3.15pm,3.51pm,5.13pm, 5.31pm, 10.35pm, 10.53pm}; or
3. {3.15pm,3.51pm,5.13pm,5.31pm,10.35pm, 10.53pm,1.35am,1.53am}; or
4. {5.13pm,5.31pm,10.35pm,10.53pm,1.35am, 1.53am, 3.15am, 3.51am}; or
5. {10.35pm,10.53pm,1.35am,1.53am,3.15am, 3.51am,5.13am,5.31am}; or
6. {1.35am,1.53am,3.15am,3.51am,5.13am, 5.31am,10.35am,10.53am}; or
7. {3.15am,3.51am,5.13am,5.31am,10.35am, 10.53am,1.35pm,1.53pm}; or
8. {5.13am,5.31am,10.35am,10.53am,1.35pm, 1.53pm,3.15pm,3.51pm}.

Depending on one's daily routine, any one of the eight possibilities can be chosen as prayer times. They provide the opportunity to those who yearn for a continual praise of the Father even during working hours.

The first set, in particular, is convenient for most office workers who normally work between 8am and 5pm. The prayer times do not take away family or personal time and rest after and before work.

For those who have time or the determination to cover the entire 24-hour period, listed below are the options for prayer times. There are 16 discrete times. The number 16 means "God's love".

Either:

1. {10.35am,10.53am,1.35pm,1.53pm,3.15pm, 3.51pm, 5.13pm, 5.31pm}; and
2. {10.35pm,10.53pm,1.35am,1.53am,3.15am, 3.51am,5.13am,5.31am};

Or:

1. {1.35pm,1.53pm,3.15pm,3.51pm,5.13pm, 5.31pm, 10.35pm, 10.53pm}; and
2. {1.35am,1.53am,3.15am,3.51am,5.13am, 5.31am,10.35am,10.53am};

Or:

1. {3.15pm,3.51pm,5.13pm,5.31pm,10.35pm,10.53pm,1.35am,1.53am}; and
2. {3.15am,3.51am,5.13am,5.31am,10.35am, 10.53am,1.35pm,1.53pm};

Or:

1. {5.13pm,5.31pm,10.35pm,10.53pm,1.35am, 1.53am, 3.15am, 3.51am}; and
2. {5.13am,5.31am,10.35am,10.53am,1.35pm,1.53pm,3.15pm,3.51pm}.

We will now show that the above times are also appropriate times to say our own personal prayers to accompany the Lord's Prayer. One of our first substantial results (Corollary 3.5[16]) says when we pray the Lord's Prayer, the grace and peace of the Father are multiplied unto us by the power of the Holy Spirit. Now in Philippians 4: 6 – 7, we read (NIV):

> [6] Do not be anxious about anything, but in every situation, by prayer and petition, with thanksgiving, present your requests to God. [7] And the peace of God, which transcends all understanding, will guard your hearts and your minds in Christ Jesus.

The word "thanksgiving" is, according to Strong's Concordance, is a cognate of the Greek adjective εὐχάριστος, *eucharistos*, that means "*grace*-ful (*thankful*) for God's *grace*". The above verses from Philippians exhort us to be thankful for God's grace whilst presenting our requests, by prayer and petition, befitting every situation, in order to experience the peace of God. Now, it is through the Lord's Prayer that we receive grace and peace. Therefore, at the prayer times in set *D*, we can accompany the Lord's Prayer with our own personal prayers of thanksgiving and requests. Then it is God's peace,

---

[16] **Corollary 3.5** *When we pray the Lord's Prayer, the grace and peace we receive from the Father is multiplied or increased by the power of the Holy Spirit.*

multiplied unto us by the power of the Holy Spirit (Corollary 3.5) that will guard our hearts and minds in union with Jesus Christ (Philippians 4:7). Corollary 3.5 and Philippians 4:6 – 7 therefore give us the following result:

Theorem 7.1 *At the times given in set D we can accompany the Lord's Prayer with our own personal prayers of thanksgiving and requests.*

Theorem 7.1, in effect, gives us the method of how to pray. At precisely the times we have chosen to pray (chosen from one of the eight options), we can start with thanksgiving and petitions, and then end with the Lord's Prayer. Or we can start with the Lord's Prayer and end with our personal prayer in the name of Jesus Christ. An earnest communication with God is necessary. Though short (the Lord's Prayer can be prayed within one minute), it must be prayed earnestly. One can pray silently as an individual, knowing that one is joined in the spirit by others at the same time at different locations. Teach others and children to do the same, for in that way, we align ourselves to the command of God in Deuteronomy 30:2, keeping in mind that The Lord's Prayer is a commandment of the Father to be prayed at times acceptable to the Father:

> [2] and you return to the LORD your God and obey His voice, according to all that I command you today, you and your children, with all your heart and with all your soul,

There are some intriguing properties of the prayer times. Let us consider the set D which is representative of all other options. Ignoring the am/pm for now, we see the following durations of pauses between prayer times:

- 10.35 – 10.53: duration 18 minutes, 18 means "oppression or bondage"
- 01.35 – 01.53 : duration 18 minutes, 18 means "oppression or bondage"
- 03.15 – 03.51: duration 36 minutes, 36 means "adversary or enemy"
- 05.13 – 05.31: duration 18 minutes, 18 means "oppression or bondage"

As we read in Chapter 6, the prayer times reflects the Agony of Jesus Christ on the Cross. He was oppressed by His enemies. Since Jesus died for us, these conversely reflect the removal of oppression and adversary from our lives when we pray the Lord's Prayer.

We have also the following durations between prayers:

- 10.53 – 01.35: duration 2 hours 42 minutes, or 162 minutes;
- 01.53 – 03.15: duration 1 hour 22 minutes, or 82 minutes;
- 03.51 – 05.13: duration 1 hour 22 minutes, or 82 minutes.

Let us look at the numbers 162 and 82 separately.

The 162nd prime number is 953, which brings up exactly two verses in the Bible with gematria 953 (Table 7.1 below, and Tables 7.2 and 7.3 in the Appendix for the gematriot):

Table 7.1: Verses with gematria 953

Text	Verse
And Moses and Aaron came before the tabernacle of the congregation.	Numbers 16:43
Commit thy way unto the LORD; trust also in him; and he shall bring *it* to pass.	Psalm 37:5

Incredibly, the first verse (Numbers 16:43) corresponds exactly to what is represented by the number 36, namely, "adversary or enemy". Numbers 16:43 is telling us that the enemies of Moses and Aaron – and therefore of God – are about to be destroyed by God for their rebellion against Moses and Aaron.

> ^{42} Now it happened, when the congregation had gathered against Moses and Aaron, that they turned toward the tabernacle of meeting; and suddenly the cloud covered it, and the glory of the LORD appeared. ^{43} <u>Then Moses and Aaron came before the tabernacle of meeting.</u> ^{44} And the LORD spoke to Moses, saying, ^{45} "Get away from among this congregation, that I may consume them in a moment."

Rather than turning away from God and risk calamity, acknowledge Him. Indeed, this is the theme of Psalm 37, the first five verses of which is given below:

> ¹Do not fret because of evildoers, Nor be envious of the workers of iniquity. ²For they shall soon be cut down like the grass, And wither as the green herb. ³Trust in the LORD, and do good; Dwell in the land, and feed on His faithfulness. ⁴Delight yourself also in the LORD, And He shall give you the desires of your heart. ⁵<u>Commit your way to the LORD, Trust also in Him, And He shall bring *it* to pass.</u> ⁶He shall bring forth your righteousness as the light, And your justice as the noonday.

Psalm 37:5 reminds us of Noah, whose commitment and trust in the Lord made him a patriarch of the Bible, a seminal head of the human race, someone who found grace in the eyes of the Lord among the wicked, the enemies of God (Genesis 6:8):

> ⁵Then the LORD saw that the wickedness of man *was* great in the earth, and *that* every intent of the thoughts of his heart *was* only evil continually. ⁶And the LORD was sorry that He had made man on the earth, and He was grieved in His heart. ⁷So the LORD said, "I will destroy man whom I have created from the face of the earth, both man and beast, creeping thing and birds of the air, for I am sorry that I have made them." ⁸<u>But Noah found grace in the eyes of the LORD</u>.

The gematria of Genesis 6:8 is 421 (Table 7.4).

Table 7.4 Gematria of Genesis 6:8 is 421

Transliteration	English	Hebrew	Gematria
wə-nō-aḥ	But Noah	ונח	64
mā-ṣā	found	מצא	131
ḥên	grace	חן	58
bə-'ê-nê	in the eyes	בעיני	142
Yah-weh.	of the LORD	יהוה	26
			421

It is the 82nd prime number, corresponding precisely to the second number – 82 minutes, the duration between the prayer times of the Lord's Prayer! In fact, Genesis 6:8 is the only verse in the Bible with gematria 421. This number continues with the overarching theme in Numbers 16 and Psalm 37 dealing with the enemies of God and God's judgment of them.

In summary, the Lord's Prayer, prayed earnestly in accordance with the 8 specific times when the Father lends His ears to us, daily clears every obstacle – spiritual and physical – in our lives whilst getting us closer and closer to His Son and our Savior, Jesus Christ. Because it is the daily proclamation of our faith in Him, daily *we walk by faith, not by sight* (2 Corinthians 5:7).

CHAPTER 8: The Lord's Prayer is the Covenant of Jesus Christ

We end boldly with the statement that the Lord's Prayer is a *covenant* that our Lord and Savior Jesus Christ made with us!

By Brown-Driver-Briggs (BDB), a covenant (תירב, *berith*) between God and man can be a *divine constitution or ordinance with signs and pledges*. A definition of *ordinance* in Collins English Dictionary is *an established or prescribed practice or usage, especially a religious rite.*

Now from Chapters 5 and 6 we recall two results:

Corollary 5.3 *It is the will of the Father that we should pray the Lord's Prayer at the times acceptable to Him.*

Theorem 6.4 *The Lord's Prayer is to be prayed daily at the times given in set D.*

Praying the Lord's Prayer at the times acceptable to the Father is a prescribed practice of religious worship. So Theorem 6.4 shows that the Lord's Prayer, to be prayed daily, is an ordinance. The Lord's Prayer is given to us by Jesus Christ. In John 12:49 (New KJV), we read:

> [49] For I have not spoken on My own *authority;* but the Father who sent Me gave Me a command, what I should say and what I should speak.

Thus it is clear that the Lord's Prayer is a commandment or divine command of the Father. Therefore, praying it daily is a divine ordinance. From this perspective, Theorem 6.4 shows that the Lord's Prayer is a divine ordinance.

We shall now show that the Lord's Prayer is a covenant, that is, it is a divine ordinance with *signs* and *pledges*.

Pledges
We have already unraveled the pledges of the Father and the Son in the Lord's Prayer. They are:

- **Corollary 3.5** *When we pray the Lord's Prayer, the grace and peace we receive from the Father is multiplied or increased by the power of the Holy Spirit.*
- **Corollary 3.10** *When we pray the Lord's Prayer, we are led by the Holy Spirit to become more like Jesus Christ.*
- **Theorem 3.11** *When we pray the Lord's Prayer, the grace we receive out of the fullness of Jesus Christ is multiplied or increased by the power of the Holy Spirit.*

Signs
Here we bring up Isaiah 59, an intriguing chapter that records in some detail – from verses 1 to 15 – the sins and transgressions of the ancient kingdoms of Judah and Israel, much to the displeasure of God (Isaiah 59:15, NIV):

> [15]Truth is nowhere to be found, and whoever shuns evil becomes a prey. The LORD looked and was displeased that there was no justice.

Verses 16 and 19 record God's dismay at the absence of a single righteous man who could intervene on His behalf, and His decision to intervene Himself and what He would do and its impact:

> [16]He saw that there was no one, he was appalled that there was no one to intervene; so his own arm achieved salvation for him, and his own righteousness sustained him. [17] He put on righteousness as his breastplate, and the helmet of salvation on his head; he put on the garments of vengeance and wrapped himself in zeal as in a cloak. [18] According to what they have done, so will he repay wrath to his enemies and retribution to his foes; he will repay the islands their due. [19] From the west, people will fear the name of the LORD, and from the rising of the sun, they will revere his glory. For he will come like a pent-up flood that the breath of the LORD drives along.

However, in verse 20, Isaiah wrote of a Redeemer who would save those who turn away from their sins:

> [20]"The Redeemer will come to Zion, to those in Jacob who repent of their sins, declares the LORD.

And with those who repent of their sins, God made a covenant with them, as recorded in the final two verses of Isaiah 59:

[21] "As for me, this is my covenant with them," says the LORD. "My Spirit, who is on you, will not depart from you, and my words that I have put in your mouth will always be on your lips, on the lips of your children and on the lips of their descendants—from this time on and forever," says the LORD.

The covenant cannot be more obvious!

Firstly, in John 14: 15 – 18, Jesus Christ referred directly to God's Spirit who would dwell in us forever (New KJV):

> [15] "If you love Me, keep My commandments. [16] And I will pray the Father, and He will give you another Helper, that He may abide with you forever— [17] the Spirit of truth, whom the world cannot receive, because it neither sees Him nor knows Him; but you know Him, for He dwells with you and will be in you. [18] I will not leave you orphans; I will come to you.

And in verses 25 and 26, we read that God's Spirit would ensure that we remember all the words of Jesus Christ and therefore of God:

> [25] "These things I have spoken to you while being present with you. [26] But the Helper, the Holy Spirit, whom the Father will send in My name, He will teach you all things, and bring to your remembrance all things that I said to you.

Indeed we infer from verse 26 that that the Redeemer of Isaiah 59 is but Jesus Christ!

Secondly, Jesus Christ taught us His prayer. The words of the Lord's Prayer are "put in (our) mouth" by Him, and with the help of the Holy Spirit, they "will always be on (our) lips, on the lips of (our) children and on the lips of (our) descendants – from this time on and forever."

Finally, we show that there is a "sign" of the covenant in Isaiah 59:21. The gematriot of the words in the first sentence (KJV) *As for me, this is my covenant with them, saith the Lord* are given in Table 8.1. The total gematria is 1811.

Table 8.1 The total gematria of the first sentence of Isaiah 59:21 is 1811

Translit.	Hebrew	English	Gematria
wa-'ă-nî,	ואני	and As for me	67
zō<u>t</u>	זאת	this	408
bə-rî-<u>t</u>î	בריתי	[is] my covenant	622
'ō-w-<u>t</u>ām	אותם	with them	447
'ā-mar	אמר	said	241
Yah-weh,	יהוה	the LORD	26
			1811

Surprisingly, the number 1811 is also the number of the very object represented by the numbers in set *D* that is the prayer pattern of the Lord's Prayer – the cross itself on which Jesus Christ was crucified! The number 1811 is the gematria of the word "cross" in John 19:25, given in context below (New KJV):

> [25] Now there stood by the <u>cross</u> of Jesus His mother, and His mother's sister, Mary the wife of Clopas, and Mary Magdalene. [26] When Jesus therefore saw His mother, and the disciple whom He loved standing by, He said to His mother, "Woman, behold your son!" [27] Then He said to the disciple, "Behold your mother!" And from that hour that disciple took her to his own home. [28] After this, Jesus, knowing that all things were now accomplished, that the Scripture might be fulfilled, said, "I thirst!" [29] Now a vessel full of sour wine was sitting there; and they filled a sponge with sour wine, put it on hyssop, and put it to His mouth. [30] So when Jesus had received the sour wine, He said, "It is finished!" And bowing His head, He gave up His spirit.

Who would have thought that biblical mathematics will point to the cross of Jesus Christ as the sign of the covenant of the Redeemer of Isaiah 59 – the sign of our salvation!

We arrive therefore at only one conclusion: *Jesus Christ has made a covenant with those who see and believe in Him, and that covenant is His very own prayer!*

Theorem 8.1 *The Lord's Prayer is the covenant of the Redeemer of Isaiah 59, Jesus Christ.*

Corollary 8.2 (Sign of the Covenant) *The Cross of Jesus Christ is the sign of the covenant of the Redeemer of Isaiah 59.*

In conclusion, the prayer pattern of the Lord's Prayer is a continual reminder of the message of the cross, for it daily focuses the attention on and belief in Christ's death on the cross, which is central to the Christian faith, as we read in 1 Peter 2:24 – 25 (ERV):

> [24] Christ carried our sins in his body on the cross. He did this so that we would stop living for sin and live for what is right. By his wounds you were healed. [25] You were like sheep that went the wrong way. But now you have come back to the Shepherd and Protector of your lives.

Hence the practical impact of following the prayer pattern is world-wide salvation.

Indeed, assume that at the indicated times the Lord's Prayer begins in Fiji, which, by virtue of its Coordinated Universal Time (UTC+12) and its coordinate on the 180th meridian (16°9'S 180°E) is the first country to see the Sun rises. Then like a tsunami, emanating from Fiji, the Lord's Prayer reverberates around the world. As the Lord's Prayer follows the Sun into new time zones, believers take over the prayer baton in the global glorification of the Father and the Son, continually, unending. Consequently, the lives of billions of people across the planet will be uplifted and changed for the better, a fitting preparation of humankind for the return of Jesus Christ (Luke 12:35 – 36, NIV):

> [35] "Be dressed ready for service and keep your lamps burning, [36] like servants waiting for their master to

return from a wedding banquet, so that when he comes and knocks they can immediately open the door for him.

With the Lord's Prayer constantly on our lips, our eyes are bound to be fixed continually on Jesus Christ who is the Author and Finisher of our faith (Hebrews 12:2).

APPENDIX
Hebrew Gematriot

There are 22 letters of the Hebrew alphabet (Table A1). They are all consonants.

Table A1: Gematriot of the Hebrew letters

Position	Name	Hebrew Letter	Gematria Standard Value
1	Aleph	א	1
2	Bet	ב	2
3	Gimel	ג	3
4	Dalet	ד	4
5	Heh	ה	5
6	Vav	ו	6
7	Zayin	ז	7
8	Chet	ח	8
9	Tet	ט	9
10	Yod	י	10
11	Kalph	כ	20

12	Lamed	ל	30
13	Mem	מ	40
14	Nun	נ	50
15	Samekh	ס	60
16	Ayin	ע	70
17	Peh	פ	80
18	Tsaddi	צ	90
19	Quph	ק	100
20	Resh	ר	200
21	Shin	ש	300
22	Tav	ת	400

There are five Hebrew letters that are written differently, but still pronounced the same, when they appear at the end of a word. They are known as *sofit* letters (Table A2).

Table A2: Gematrias of the Hebrew sofit letters

Position	Name	Letter	Gematria Standard Value	Sofit Letter	Gematria Sofit Value
11	Kalph	כ	20	ך	500
13	Mem	מ	40	ם	600
14	Nun	נ	50	ן	700
17	Peh	פ	80	ף	800
18	Tsaddi	צ	90	ץ	900

Greek Gematriot

The Greek alphabet is straightforward (Table A3):

Table A3: Gematriot of the Greek letters

Position	Name	Greek Letter	Gematria Standard Value
1	Alpha	Α, α	1
2	Beta	Β, β	2
3	Gamma	Γ, γ	3
4	Delta	Δ, δ	4
5	Epsilon	Ε, ε	5
6	Zeta	Ζ, ζ	7
7	Eta	Η, η	8

8	Theta	Θ, θ	9
9	Iota	Ι, ι	10
10	Kappa	Κ, κ	20
11	Lambda	Λ, λ	30
12	Mu	Μ, μ	40
13	Nu	Ν, ν	50
14	Xi	Ξ, ξ	60
15	Omicron	Ο, ο	70
16	Pi	Π, π	80
17	Rho	Ρ, ρ	100
18	Sigma	Σ, σ, ς	200
19	Tau	Τ, τ	300
20	Upsilon	Υ, υ	400
21	Phi	Φ, φ	500
22	Chi	Χ, χ	600
23	Psi	Ψ, ψ	700
24	Omega	Ω, ω	800

Tables of Gematriot

Tables for Chapter 1

Table 1.2: Gematria of Isaiah 1:19 is 2076

Transliteration	English	Hebrew	Gematria
'im-	If	אם	41
tō-ḇū	you be willing	תאבו	409
ū-šə-ma'-tem;	and obedient	ושמעתם	856
ṭūḇ	the good	טוב	17
hā-'ā-reṣ	of the land	הארץ	296
tō-ḵê-lū.	you shall eat	תאכלו	457
			2076

Table 1.3: Gematria of James 2:17 is 4730

Transliteration	English	Greek	Gematria
houtōs	So	ουτω	1570
kai	also	και	31
hē	-	η	8
pistis	faith,	πιστις	800
ean	if	εαν	56
mē	not	μη	48
echē	it has	εργα	109
erga	works,	εχηι	623
nekra	dead	νεκρα	176
estin	is.	εστι	515
kath'	by	καθ	30
heautēn	itself.	εαυτην	764
			4730

Table 1.4: Gematria of 1 Kings 17:4 is 3004

Transliteration	English	Hebrew	Gematria
wə-hā-yāh	And it shall be	והיה	26
mê-han-na-ḥal	of the brook	מהנחל	133
tiš-teh;	[that] you shall drink	תשתה	1105
wə-'eṯ-	and	ואת	407
hā-'ō-rə-ḇîm	the ravens	הערבים	327
ṣiw-wî-ṯî,	I have commanded	צויתי	516
lə-ḵal-kel-ḵā	to provide you	לכלכלך	150
šām.	there	שם	340
			3004

Tables for Chapter 3

Table 3.2: Gematria of Mark 14:19 is 8951

Transliteration	English	Greek	Gematria
kath'	every	καθ	30
hēmeran	day	ημεραν	204
ēmēn	I was	ημην	106
pros	with	προς	450
hymas	you	υμας	641
en	in	εν	55
tō	the	τωι	1110
hierō	temple	ιερωι	925
didaskōn	teaching,	διδασκων	1089
kai	and	και	31
ouk	not	ουκ	490
ekratēsate	you did seize	εκρατησατε	940

me	me.	με	45
all'	but [it is]	αλλ	61
hina	that	ινα	61
plērōthōsin	might be fulfilled	πληρωθωσιν	2087
hai	the	αι	11
graphai	Scriptures.	γραφαι	615
			8951

Sovereignty of the Father

Table 3.3: Gematria of Romans 9:18 is 1867

Transliteration	English	Greek	Gematria
ara	So	αρα	102
oun	then	ουν	520
hon	to whom	ον	120
thelei	he wants,	θελει	59
eleei	he shows mercy,	ελεει	55
hon	whom	ον	120
de	moreover	δε	9
thelei	he wants,	θελει	59
sklērynei	he hardens.	σκληρυνει	823
			1867

Acknowledgment of the Father

Table 3.4: Gematria of Luke 11:2 (shaded) is 13,326

Transliteration	English	Greek	Gematria
eipen	he said	ειπε	
de	moreover	δε	
autois	to them,	αυτοις	
Hotan	When	{οταν	
proseuchēsthe	you pray	προσευχησθε	

legete	say,	λεγετε	
Pater	Father	πατερ	486
hēmōn	of us,	ημων	898
ho	who [is]	o	70
en	in	εν	55
tois	-	τοις	580
ouranois	heaven,	ουρανοις	901
hagiasthētō	hallowed be	αγιασθητω	1332
to	the	το	370
onoma	name	ονομα	231
sou	of you;	σου	670
elthetō	let come	ελθετω	1149
hē	the	η	8
basileia	kingdom	βασιλεια	259
sou	of you;	σου	670
Genēthētō	let be done	γενηθητω	1183
to	the	το	370
thelēma	will	θελημα	93
sou	of you,	σου	670
hōs	as	ως	1000
en	in	εν	55
ouranō	heaven,	ουρανωι	1431
kai	[so] also	και	31
epi	upon	επι	95
tēs	the	της	508
gēs	earth.	γης	211
			13326

Table 3.5: Gematria of Mark 10:34 is 13,326

Transliteration	English	Greek	Gematria
kai	And	και	31
empaixousin	they shall mock	εμπαιξουσιν	926
autō	him,	αυτωι	1511
kai	and	και	31
mastigōsousin	shall scourge	μαστιγωσουσιν	2284
auton	him,	αυτον	821
kai	and	και	31
emptusousin	shall spit upon	εμπτυσουσιν	1755
autō	him,	αυτωι	1511
kai	and	και	31
apoktenousin	shall kill	αποκτενουσιν	1256
auton	him,	αυτον	821
kai	and	και	31
tē	the	τηι	318
tritē	third	τριτηι	728
hēmeras	day	ημεραι	164
anastēsetai	he shall rise again	αναστησεται	1076
			13326

Becoming More Like Jesus Christ

Table 3.7: Gematria of 2 Corinthians 2:18 is 13,457

Transliteration	English	Greek	Gematria
hēmeis	we	ημεις	263
de	moreover	δε	9
pantes	all,	παντες	636
anakekalymmenō	having been unveiled	ανακεκαλυμμενωι	1473
prosōpō	in face,	προσωπωι	2140
tēn	the	την	358
doxan	glory	δοξαν	185
Kyriou	of [the] Lord	κυριου	1000
katoptrizomenoi	beholding as in a mirror,	κατοπτριζομενοι	1133
tēn	the	την	358
autēn	same	αυτην	759
eikona	image	εικονα	156
metamorphoumetha	are being transformed into,	μεταμορφουμεθα	1581
apo	from	απο	151
doxēs	glory	δοξης	342
eis	to	εις	215
doxan	glory,	δοξαν	185
kathaper	even as	καθαπερ	216
apo	from	απο	151
Kyriou	[the] Lord,	κυριου	1000
Pneumatos	[the] Spirit.	πνευματος	1146
			13457

Tables for Chapter 6

Table 6.2: Gematria of Zechariah 14:7 is 1487

Translit	English	Hebrew	Gematria
wə-hā-yāh	But it shall be	והיה	26
yō-wm-	day	יום	56
'e-ḥāḏ,	one	אחד	13
hū	that	הוא	12
yiw-wā-ḏa'	shall be known	יודע	90
Yah-weh	to the LORD	ליהוה	56
lō-	not	לא	31
yō-wm	day	יום	56
wə-lō-	nor	ולא	37
lā-yə-lāh;	night	לילה	75
wə-hā-yāh	but it shall come to pass	והיה	26
lə-'êṯ-	time	לעת	500
'e-reḇ	[that] at evening	ערב	272
yih-yeh-	it shall be	יהיה	30
'ō-wr.	light	אור	207
			1487

Tables for Chapter 7

Table 7.2: Gematria of Numbers 16:43 is 953

Transliteration	English	Hebrew	Gematria
way-yā-ḇō	And came	ויבא	19
mō-šeh	Moses	משה	345
wə-'a-hă-rōn,	and Aaron	ואהרן	262
'el-	before	אל	31
pə-nê	the face of	פני	140
'ō-hel	the tent	אהל	36
mō-w-'êḏ.	of meeting	מועד	120
			953

Table 7.3: Gematria of Psalm 37:5 is 953

Transliteration	English	Hebrew	Gematria
gō-wl	Commit	גול	39
'al-	to	על	100
Yah-weh	the LORD	יהוה	26
dar-ke-ḵā;	Your way	דרכך	244
ū-ḇə-ṭaḥ	and trust	ובטח	25
'ā-lāw,	also in him	עליו	116
wə-hū	and he [it]	והוא	18
ya-'ă-śeh.	shall bring to pass	יעשה	385
			953

Isaiah 53

How do we know that Jesus was the suffering Messiah of Isaiah 53? In John 1:1 – 2, we read:

> [1] In the beginning was the Word, and the Word was with God, and the Word was God. [2] He was with God in the beginning.

And in verse 9, we read:

> [14] The Word became flesh and made his dwelling among us. We have seen his glory, the glory of the one and only Son, who came from the Father, full of grace and truth.

Thus, Jesus is the Word, literally the written message of God. If so, the number of words, and therefore the number of characters or letters of the original Hebrew text in Isaiah 53 must refer to Him. Now, if we go back a chapter to Isaiah 52, there is a sudden deflection of the thoughts of the author starting from verse 13 to the end of chapter, verse 15. From verses 1 to 12, the subject is Zion. From verse 13, the subject is the suffering and glory of a servant of God. Biblical scholars accept that the same servant is referred to in Isaiah 53. Table A4 below gives the number of Hebrew words and letters in each verse from Isaiah 52:13 till the end of Isaiah 53:

Table A4: Taken from the original Hebrew text

Chapter	Verse	Number of Hebrew words	Number of Hebrew letters
Isaiah 52	13	7	27
	14	11	43
	15	18	64
Isaiah 53	1	8	32
	2	15	59
	3	13	57
	4	12	55
	5	11	54
	6	12	47
	7	16	61
	8	15	54
	9	14	48
	10	16	61
	11	12	52
	12	22	86
		202	**800**

We shall show now that the numeric 800 groups two bible verses – a verse from the Old Testament and a verse from the New Testament – under the same theme, namely, "savior".

In the Old Testament, 800 is the gematria of the words "thy salvation" – the transliteration of which is "yə·šu·'ā·têḵ", which, astonishingly according to Strong's Concordance, occurs only once in the Bible! It is in in Psalm 35:3:

¹ Plead my cause, O LORD, with them that strive with me: fight against them that fight against me. ² Take hold of shield and buckler, and stand up for mine help. ³ Draw out also the spear, and stop the way against them that persecute me: say unto my soul, I am <u>thy salvation.</u>

The Hebrew for "thy salvation" is ישועתך, *yeshuatech*, which comes from the noun ישועה, *yeshuah*, meaning "salvation". Its verb is ישע, *yasha*, that means "to deliver."

In the New Testament, the numeric 800 is the gematria of word "world" – κοσμου, *kosmou* – in the verse 1 John 4:14:

> ¹⁴ And we have seen and testify that the Father has sent his Son to be the Savior of the <u>world</u>.

The Lord and Savior of Psalm 35 is the Savior of the World in 1 John 4!

www.ingramcontent.com/pod-product-compliance
Lightning Source LLC
Chambersburg PA
CBHW031434210526
45464CB00005B/2205